WEB BASED
CORPORATE INSTITUTES

WEB BASED
CORPORATE INSTITUTES

A Solution for Unfinished Defense Industry Acquisitions

T. H. Henning

iUniverse, Inc.

New York Lincoln Shanghai

Web Based Corporate Institutes
A Solution for Unfinished Defense Industry Acquisitions

iUniverse books may be ordered through booksellers or by contacting:

iUniverse
2021 Pine Lake Road, Suite 100
Lincoln, NE 68512
www.iuniverse.com
1-800-Authors (1-800-288-4677)

ISBN-13: 978-0-595-38883-7 (pbk)
ISBN-13: 978-0-595-83262-0 (ebk)
ISBN-10: 0-595-38883-3 (pbk)
ISBN-10: 0-595-83262-8 (ebk)

Printed in the United States of America

C O N T E N T S

List of Tables

Acknowledgements

My wife, Cheryl, provided recommendations and improvements to the manuscript and motivated me to keep writing. Dr. Zachary Thomas Henning provided editorial comments and improvements to the manuscript. Mr. Chris Schaft provided valuable insights from his experience as a telecommunications and business services management consultant. I've relied on the experience gained from the various corporations that I've worked for over the years. It was an honor and a privilege for me to work with many talented training and development professionals at IBM, Loral, Lockheed Martin, The Analytic Sciences Corporation, and Veridian.

INTRODUCTION

▼

The objective of *Web Based Corporate Institutes* is to share the author's experience in assessing how the formation and use of web based corporate institutes has helped some defense corporations integrate different corporate cultures following corporate acquisitions or mergers. The analysis captures the author's experience at IBM, Loral, Lockheed Martin, The Analytical Sciences Corporation (TASC) which is now a division of Northrop Grumman, and Veridian which has been acquired by General Dynamics.

These corporations had very different corporate cultures and different approaches to managing corporate acquisitions. Each corporation developed its own corporate culture made up of behavioral patterns and beliefs that were generally accepted by the employees of that corporation. When one corporation acquired another corporation or merged with another corporation, a multicultural corporation was formed. The two or more corporate cultures now under one corporate headquarters often had different policies, practices, processes, and values. While they shared customers in the same industry, there were many problems that arose for the acquiring corporation such as differing compensation systems, personnel policies, mission integration strategies, and differing approaches to market and customer segmentation. The process for acquiring another corporation is well understood but what happens after the merger and acquisition has taken place is often ambiguous leaving lower echelon managers and employees on their own to deal with multicultural corporate differences. Few corporations have smoothly integrated their operations following an acquisition, but there have been some successes in the defense industry. Some defense corporations have developed web based corporate institutes and successfully applied organization development techniques to integrate corporate cultures after acquisitions have taken place. Most attempts, however, are still works in progress. Some have formed corporate universities to provide employee training and development programs. The term "corporate university," in my opinion, is not the correct title for this kind of corporate organization. I believe the term corporate institute, which focuses on the specialized nature of corporate training, rather than "university," which has a broader focus, is the most appropriate term for use in business. The role of an academic university, which is often made up of several colleges, is to provide general higher education and professional education to the public. Some public and private university faculty may find the term corporate university objectionable and view it as a corporate attempt to compete for the education mission assigned to their institutions. Additionally, state departments of higher education often have licensing regulations applicable to forming universities therefore laws of every state must be considered if a corporation decides to develop a corporate university. An institute, on the other hand, has a specific audience and a specialized role to provide technical, business, and professional development to a specific audience. This role can be expanded to include the organization development techniques that can integrate corporate cultures. In recent years, establishing a web based corporate institute has led to improved development of the organization, the integration of acquired businesses, and improved centralized training and development.

Advancements in Internet technologies, common widespread use of corporate intranets and the World Wide Web, and the development of web based information and knowledge management concepts over the last decade have made it possible for people to access and use knowledge and information from anywhere in the corporation and anywhere in the world. When corporations merge, these capabilities can be harnessed to better integrate the different corporate cultures and to streamline disparate policies and practices in order to develop a more harmonious flow of work and improve customer service. The development of corporate intranets that have firewalls to allow only corporate employees access to information stored there has provided an opportunity for corporations to unify their business. By storing corporate information on the intranet for employees to access and use and communicating frequently with employees, new corporate acquisitions can be successfully blended into a new corporate culture. The technology is here today that will enable executives to optimize web based resources and accelerate the integration of new corporate cultures into a more unified corporation.

Corporate cultures can be loosely defined as the learned and practiced behavior patterns of a group of people. Since groups of people learn different patterns of behavior, multi cultures are formed in corporations. Not all corporate cultural behaviors are appreciated by different corporations, so there are basically three choices managers have in coping with different corporate cultures. First, they can integrate the cultures by suppressing behavior that is not acceptable and encouraging new behavior. Second, they can segregate the cultures into like groups and leave each group to its own culture. Finally they can ignore the issue, plunder the acquired company for assets and layoff the employees of the acquired business. In defense corporations and, I suspect, other corporations that grow by acquiring other corporations, the third option often results in low morale, tribal like corporate cultural wars between organization functions, and clashes between managers and employees. Dealing with the fall out of unfinished acquisitions often preoccupies management in many defense corporations and diverts attention from normal business.

In my experience, defense industry corporations which developed web based corporate institutes tended to have better success integrating new corporate acquisitions into the company. Much of their success would be attributed to top management's focus on web based information, training, and organization development initiatives designed to alter previously accepted corporate behaviors and beliefs. This kind of social engineering was aimed directly at achieving a single corporate culture that was readily identifiable by customers and distinguishable from other competitors. More and more corporations are growing by acquiring other companies. This is particularly true in the defense industry. The audience that may find this book useful includes but isn't limited to corporate chief executive officers, corporate training and human resource development managers, and managers and professionals tasked with organization development.

T.H. Henning
Haymarket, Virginia
March 31, 2006

▼

CHAPTER 1

THE CHALLENGE OF MANAGING UNFINISHED CORPORATE MERGERS

A multicultural defense corporation is formed after one defense company acquires another defense company. Each corporation has its own corporate culture, policies, practices, and values. The struggle that ensues after the corporate acquisition is a conflict of management styles. It is pointless to impose one company's culture on an acquired business. It is better to understand the strengths and weaknesses of both corporate cultures and forge ahead with a new business culture.

The defense industry has been characterized by divestitures, mergers, and acquisitions for decades. As corporations grow by acquiring other corporations, they are beset with clashes of corporate cultures. A corporation forms a culture by training and developing employees to believe in certain values held by that corporation. By training employees to follow certain policies, practices, and adhering to certain beliefs, a corporate culture is formed over time. The culture is reinforced when the company becomes successful in the marketplace and employees benefit from the training they've received. The company's compensation plan rewards successful accomplishments thereby reinforcing the corporation's culture. Each corporation is unique in the sense that they believe their corporate culture is superior to that of other corporations. When one corporation acquires another, or divests a portion of its corporation to another company, the result is an immediate clash of processes, practices, and beliefs.

It has been my experience to work for several defense industry corporations that acquired and divested major business divisions to other corporations and to observe and participate in the tribal-like warfare that followed the corporate merger. It was also my experience to work for major corporations that acquired other corporations and successfully integrated them into a single corporate enterprise before they, too, were acquired by other major defense corporations. It has been my experience that *the future of corporations is written in the hearts and minds of their employees therefore the key to integrating multicultural defense corporations is how they train and develop people.*

All over the world defense corporations are increasingly turning to corporate trainers not only to develop knowledge, skills, and attributes of people who they hope will lead their companies into a profitable future, but also to help integrate people from different corporate cultures into a single corporate culture. Billions of dollars are spent annually in the U.S. on training and development but corporations struggle to assess whether their efforts are achieving the results they had hoped for. Many employees remain skeptical of their corporate leaders and remain ill prepared to face the complexities of the ever-changing business environment. When business divestitures and acquisitions occur, usually the last item management thinks of is how employee and management behavior must be altered to adapt to the new corporation being formed. While most employees are grateful for any training sponsored by the company, many believe that the training and development investment is insufficient to keep pace with their need for continuous learning, particularly if they find themselves in a new corporation with new policies, practices, technologies, and procedures. Corporations pressured by costs and increasingly competitive market conditions do not focus as intensely as they should on human resource development. During difficult times and times of transition, training and development budgets are usually the first targets of cost reduction yet this is when training and development is most needed. Human resource programs are often seen as side shows to the main event of making a profit. A focus on increasing profits, and ignoring the social, political, economic, and productivity loss impact of poor people management all point to a lack of enlightened strategic leadership in many corporations. *This short term view is further supported by boards of directors and the marketplace itself which encourages quarter to quarter results and does not encourage strategic and long term planning.*

The complex nature of the business environment leads to complexity in preparing people to adapt to major and sudden change. Many companies don't think of succession planning, (where new visionary leadership will come from), or what kind of leadership is required in the organization after it has acquired another defense company. *New marketplace challenges, new technologies, new patterns of social interaction, cultural diversity, globalization, the high velocity pace of business, and multicultural corporate conflicts are just a few of the business environment challenges.* These forces of change impact human resource development in several ways including but not limited to:

- the need for new leadership knowledge, skills, and attributes,

- the need to keep pace with changing marketplace products and services,

- the need for employees to keep pace with and try to lead technology change in a business environment that is increasingly interdependent, and

- the need to understand how corporate cultures form, how they change as a result of corporate acquisitions, and the impact corporate culture has on the performance of individuals in the company.

NEW MARKETPLACE CHALLENGES

The marketplace today is not only local, regional, or national. It is increasingly international. Organizations partner with, and acquire firms around the world to market and service their products. Globally, new products and services are being developed and introduced for sale at a rapid pace. Marketing and sales training must struggle to keep pace with the array of products and services offered by their organization as well as competitor corporations around the world. Senior leaders must search continuously for that competitive edge that will distinguish their products and services from those offered by their competitors. A web of international supplier relationships must be understood and managed effectively.

NEW TECHNOLOGY CHALLENGES

Engineers, software developers, information technology professionals and other technical personnel are struggling to keep up with new technologies and advances in their professions. Technologies are introduced on a global scale as well as from the research labs in domestic corporations. The application of these new technologies and rush to get them to the marketplace has accelerated the need for technical training. New college and university graduates find that they need continuous life long learning to keep up with advances in their own professions and often rely on their corporations to help them. My experience is that after a corporate merger in the defense industry, there are many different kinds of new workstation computers, software, and technology infrastructure that employees must understand and utilize. New communications tools and processes must be understood to be used effectively.

NEW PATTERNS OF SOCIAL INTERACTION AND CULTURAL DIVERSITY

Demographic changes in the U.S. population as well as the populations of countries around the world show that in some countries the population is aging and in others there's an abundance of educated youthful workers. Social behavior in the workplace can be a challenge, particularly after one corporation acquires another. Some corporations spend millions of dollars on training courses which focus on workplace interpersonal interaction and behavior issues including cultural diversity, harassment, ethics, and compliance with legal statutes and regulations. More recently they've offered team building courses which encourage employees to collaborate rather than to compete inside the corporation. Getting employees to ignore age, race, and gender and focus on job specific results requires on going training.

The workforce has also changed in many countries of the world. For example, it is not uncommon for native residents in European corporations to work with people from many countries and diverse cultures. Compliance and understanding of individual and cultural differences is just part of the challenge of human resource development. Training and development programs need to be oriented to help people understand diverse religious beliefs, languages, cultural behaviors and their effects on workplace behavior. Prior to conducting international business it is the practice of many corporations to prepare their employees who travel abroad, by providing training in government requirements, legal requirements, languages, and cultural differences. When corporate mergers or acquisitions occur, this complex environment is challenged by different corporate approaches to managing those issues.

GLOBALIZATION

International business in the last century was often conducted by having plants, labs, and branches in overseas offices staffed mostly by people from the resident foreign country. The world has become more complicated since then and managers from various countries and cultures of the world lead corporations from countries different than their own. When we speak of globalization today we are referring to World Wide Web communications, world-wide partnerships, global suppliers, the outsourcing of work to foreign companies, the rapid pace of international trade, and a host of new global marketplace trends. Corporate employees need new knowledge, new skills, and new attributes to work in business today. University professors often do not have current corporate knowledge needed to be able to incorporate this new knowledge into college and university courses. *Consequently, web based corporate institutes must be designed to continue the education of corporate managers and employees.* Many major defense industry corporations are international and global in their mission and scope. Acquisitions and mergers of U.S. national and foreign defense businesses require U.S. government approval and when they occur there are additional issues to be managed.

HIGH VELOCITY PACE OF BUSINESS

With the introduction of wireless mobile computing, cell phones, the World Wide Web, the Internet, and a host of new communications technologies, the pace of business has accelerated tremendously over the past decade. Many professional employees find themselves in a stressful environment where they have little to no personal time to enjoy their personal lives. Too much work with too little time to do it is a common complaint. Human resources impacts to the rapid speed of business can result in costly mistakes, product quality issues, and customer service problems. Consider the many channels of communications that workers have today. Messages from email, regular office mail, office phone mail, cellular phone mail, web phone Internet mail, pagers, etc. bombard the employee with numerous messages every day. Mountains of messages await the employee who returns to the office after taking a few days of vacation. When corporations acquire other business units or other corporations, the pace of business sometimes adds additional stress to employees and managers and strategic plans are overwhelmed by rapid change.

MULTICULTURAL ORGANIZATION CONFLICT

And if the above challenges were not enough, corporation's today wrestle with how to streamline their corporation that has grown through acquisition of other corporations. This often means accepting new processes and practices of the acquired corporation as well as eliminating duplicate overhead costs and duplicate overhead staffs. Maintaining and sustaining high employee morale is impacted by interpersonal and inter organizational conflict which often results from the clash of corporate ways of conducting business. It was my experience that following a divestiture, merger, or business acquisition, staff functions became misaligned and out of step with operational functions, the decision making process became muddled, and conflict between staff functions like finance, human resources, marketing, and procurement frustrated operational line management. Many of these conflicts have to be resolved at the highest level of the corporation since they are often complex group to group not individual to individual disagreements.

HUMAN RESOURCES

Most CEO's and corporate leaders often turn to human resources, (HR), to integrate new corporate employees into their company. But HR is not up to the task of integrating an entire corporate culture into the mainstream of their company. Most human resource professionals have focused on their own career priorities such as keeping up with benefits changes, new labor laws, compensation, labor relations, recruiting, and employee relations. Integrating a corporation with another corporation often means dealing with different benefit programs, different compensation programs, different awards programs, and different HR policies and practices. Both corporate entities believe their programs are the best for their employees. Executive management must intervene and settle these tribal-like conflicts and decide on a case by case basis which course of action will benefit the employees and managers of the multicultural corporation. Additionally, human resource development programs often are not a top priority. Many of the human relations professionals have little or no practical experience in the operations and mainstream products and services of the corporation and lack first hand experience with the training requirements or main business operations of the corporation. I've found this to be particularly true in the defense industry where most operations require security clearances and technical skills and knowledge. Human resources personnel often do not have security clearances required for government contract work and since many of them have never worked in a defense contracting function and lack technical skills and knowledge, they often cannot relate to the technical training requirements of the organization. *Training and developing people is often delegated to the human resources department where it often becomes fragmented, poorly managed, and generally prioritized low by many senior human resources managers.* If traditional human resources functions, (HR), can be thought of as **acquiring** human resources, (recruiting, hiring, etc.), **administering** human resources, (benefits, compensation, employee relations, records administration, etc.), and **developing** human resources, (orientation, indoctrination, training, etc.), then I believe the third element of this traditional view of HR can best be done by a web based corporate institute. The web based corporate institute should take the lead in developing human resources and integrating people from the acquired corporation into the acquiring corporation. In the future, much of the training and development activity which today is labor-intensive and costly will be accomplished on-line. Much of human resource administration can also be accomplished on line and increasingly, recruiters are using on line technology to interview applicants using teleconferencing and web based conferencing technologies. Other HR administrative information can also be made available on line and can be modified as policies and practices of the new corporate entity are changed.

Adults and senior professionals have already learned a great deal and developed their own preferences for learning new information. Classroom instruction sponsored by many HR departments often doesn't take into account that instructors are dealing with adult learners. Complex social and psychological processes are at work in group settings. A lock-step "one size fits all" classroom instruction plan doesn't yield the results sought in many cases because the learning process is highly individual.

Social, psychological, and demographic factors combined with the rapid pace of technological change indicate that if a company does not train and develop its human resources, it will not retain the best people they have and will not be able to compete in the global marketplace let alone integrate different corporate cultures. Human resources organizations competing for talented new hires in the marketplace today try to characterize their corporations as having competitive compensation and benefits plans and to represent their organizations as having pleasant work environments and advancement opportunities. *When mergers or acquisitions occur, the recruiting process must reflect one corporation to prospective new hires and not offer different benefits for new employees depending on which part of the corporation they're hired by; the heritage acquired corporation or the parent acquiring corporation.* One of the appealing attractions for new employees is to be able to continue their education. Tuition reimbursement plans are often highlighted in corporate recruitment literature and they are often very different from corporation to corporation. A web based corporate institute can serve to attract and retain new employee candidates as well as re-orient employees from acquired corporations to the resources available in the combined corporations. Both of the corporate web based institutes I developed had on going relationships with colleges and universities nation-wide. Participation by the faculty from those colleges and universities in corporate institute programs helped to attract students from those universities to the corporations and the use of outside speakers and faculty led to more objective discussions of the merits of differing corporate approaches to operational issues.

TEAM BUILDING

Many corporations train teams to develop proposals and bid on potential new contract opportunities. Teams are also formed to address special organizational cross function issues such as quality, technical process improvement, etc. A web based corporate institute can facilitate team building and improve the performance of teams by providing on line team meetings using on line meeting tools. Access to on line repositories of organizational information through the web based institute's intranet links can help improve the speed and performance of teams. Several defense industry corporations I've worked for found that on line web based meetings helped save travel time to meeting locations. Their web based corporate institutes served to provide an environment that was conducive to soliciting a wide range of opinions and information on the topic the team was investigating. For example, at TASC, the team was investigating how to improve its systems engineering and software engineering processes. The corporate institute arranged on line access to the Software Engineering Institute that focuses on providing corporations with on line repositories of information and guidance on how to improve those processes. The team's final report included improvements that the corporation could make not only to systems and software engineering processes but also to quality and staff support services processes. Implementation of the team's recommendations helped the corporation achieve its goals. At Veridian the web based corporate institute established and chaired a team that was investigating how to improve project management application software tools. The web based corporate institute provided on line access to lists of project management tools available on the market, evaluation of the tools costs, and invited the participation of customers and external faculty to provide their views of project management tools at seminars and in emails. The advantage of having a web based corporate institute provide this service rather than a functional organization in the corporation is that through its supply chain and partnerships the web based corporate institute gained access to expert opinions and on line resources quickly. As a cross-functional entity located at the corporate level it could draw from many internal as well as external sources of information, and a web based corporate institute served to coordinate the flow of information for the team on an organizational level rather than on a functional level thereby opening up more channels of ideas and information.

MANAGING HUMAN RESOURCE CHALLENGES IN MULTICULTURAL DEFENSE CORPORATIONS

Itemized below are some ways defense corporations have dealt with human resource challenges and addressed behavior change in a multicultural corporate environment.

Web Based Corporate Institute

To address new and challenging human resource issues, some corporations have chosen to develop a web based corporate institute. Chapter Five describes the process of developing this different kind of corporate institute that is web based and provides all employees and managers with the ability to access the knowledge, skills, and information they need, any time, any place. A web based corporate institute can provide for knowledge and information sharing as well as facilitate collaboration between managers and employees of the corporation. *Many problems in human resources management deal with poor leadership, poor employee behavior on the job, and stress involved in job performance. The solution for many of these problems is training.* While training is not the solution to every problem, it's a start in dealing with human resource challenges.

New Ways of Using On Line Resources

Applications of distance learning and blended learning technologies offer not only an alternative to the classroom, but also a way to facilitate collaboration and information sharing throughout the corporation. In some cases, rather than taking full on line courses employees can access multiple on line courses through hyperlinks to gain insight into solving immediate problems on the job. In this example, on line courses can also be used for job performance support. Some may argue that one has to complete an entire course to obtain credit on ones education record with the corporation and there's merit in this argument. But it does not preclude creative methods and short cuts that people may take to quickly improve their performance on the job. There is an abundance of information available on the Internet and the World Wide Web. New methods should be sought to sort out the information one really needs, provide thoughtful analysis of the information, and apply the analyzed information to solving the problems the organization seeks to solve. The process begins with gathering data. From data one can gather information from sets of data. From information, one can, through analysis, gain knowledge. And gaining knowledge, one can then apply it to solve problems.

Optimize the Technology Infrastructure to Facilitate Collaboration

There are those that will argue that a web based corporate institute will not allow for social interaction, will not accommodate all types of learning or learners, and that there will always be a need for group, classroom delivery. There is some truth to this argument. *But it's my contention that the rapid advances now being made in on line simulations, virtual learning, and distance learning will allow for as much social interaction as occurs in today's classrooms and will accommodate all types of learning styles and learners.* Classroom delivery of knowledge, skills, and attributes will be increasingly viewed as not being able to keep up with the pace required by tomorrow's requirements. Trying to integrate corporate cultures in a classroom setting may cause more problems than it solves. The classroom has served us well in the 19th and 20th centuries, but the 21st century calls for a paradigm shift to on line learning of knowledge, skills, and attributes sponsored by web based corporate institutes. Collaboration on the job and formation of web enabled teams addresses some of the human resource challenges corporations face in the 21st century. Frequent web casts from the Chief Executive Officer, (CEO) discussing progress made in integrating the acquired corporations employees into their new corporate identity goes far to improve morale and assist in integrating the multicultural corporation.

Look to the Future

Looking ahead toward the future after looking back on lessons learned from the past. can help management set a vision for the multicultural corporation. A vision that takes advantage of the strengths of both corporations is highly desirable. *The trends in training and developing employees indicate that employees will be sourcing the knowledge and skills they need from a web-based source. You can see this trend in operation today. Many people now use web based information sources for gathering encyclopedic information.* They use search engines in order to find the information and knowledge they need and surf the web for information rather than go to the local library. It's faster and easier to use on line resources. In corporations, the web is used more and more frequently. Wireless technologies and access to web based resources from personal data assistants or cell phones also are ways to facilitate communications.

To optimize the flow of information and collaboration in the corporation, it has been my experience that a corporate web based institute can facilitate initiatives to improve the corporation's processes and human resource development programs. On line mentoring, coaching, and other initiatives can be a function of the corporate institute. A web based corporate institute can facilitate the introduction and application of new web based software tools as they occur in the marketplace and be a repository for corporate knowledge and lessons learned. Combining resources of the acquired and parent organization helps employees collaborate on work projects. ***The challenges of managing multicultural organizations are primarily human resource development challenges.***

LESSONS LEARNED

- The challenge of managing human resources in multicultural corporations combined with the rapid pace of change, the explosion of new knowledge, and the increasing diversity of the workforce means that there is a requirement for life long learning and more effective leadership.

- Lack of strategic planning following business acquisitions in defense corporations can lead to attrition, low morale, and potential loss of business.

- A focal point for integrating people into the parent corporation should be an on line corporate web based institute.

- The difficulty of pulling people off the job to be trained that results in loss of productivity can be offset with new technology that can provide employees with access to job performance support tools and training any time, any place.

- The management resistance to consolidating training budgets in the enterprise often means resistance to change in general. The trend in most organizations is to centralize training and development rather than to disperse it throughout the organization.

- There is generally poor top executive financial visibility into the costs of training and how they differ from other business maintenance costs.

- Executive management views training and development as a cost to the business and not an investment, yet there is a desire to see a return on investment. Management needs to clarify what constitutes an investment and what constitutes a cost. During difficult times and during times of transition more focus on training and developing the workforce is required.

- Placement of human resource development in the human resources function where it often does not receive the priority it deserves should lead to a review of where a corporate institute should report in the corporation and how web based services can improve the overall effectiveness of the corporation.

- Team building and improving team performance can be facilitated by a web based corporate institute.

- Management needs to construct new, cost effective, on-line resources accessible to all and overcome internal barriers to implement new ways of addressing old problems. Management needs to lead change rather than react to it.

- Integrating similar missions is relatively easy when one corporation acquires another, but integrating and streamlining the support functions isn't given the priority it needs to smooth the transition and can cause problems and disruptions.

CHAPTER 2

▼

STRATEGIES AND OPTIONS

> The most valuable asset a corporation possesses is its professional labor force. When acquiring another corporation the acquiring corporation must have a plan and strategy for incorporating these talented people into the new business environment. People have irreplaceable value. The knowledge, ability, and experience they bring to the corporation is an asset that requires top management attention all the time.

Most executives recognize the importance of recruiting and maintaining highly skilled and knowledgeable workers as essential to supporting their business. Many corporate leaders are investigating better strategies and options to improve corporate performance in handling the problems and issues that arise after mergers or acquisitions.

Usually corporation executives do not have a strategy to change the corporate culture. Corporations are in business to make a profit and to create wealth in a society so their most important strategy is profit optimization and market domination. Duplicate overhead costs must be avoided so following an acquisition they act to combine and consolidate staff and operational functions. Acquired corporations usually have most of the same staff functions that the acquiring corporation has, but it would be a mistake to assume that eliminating all of the acquired corporation's staff functions would necessarily result in lower overhead costs. A transition period is often part of the strategic planning process. Here are some different strategies that some corporations have found successful in developing and merging an acquired corporate asset.

DEVELOPING A CORPORATE CULTURE

Some corporations want to continuously evolve their corporate culture and some want to dramatically change it. The IBM corporate culture was formed following a change of top leadership at The Computing, Tabulating, and Recording Company (CTR) in the early 1900's.

IBM

Tom Watson Sr. had a belief that there was no saturation point to education and believed that an investment in people would benefit a corporation. When he became the general manager of CTR, he took steps to develop a new corporate culture. He renamed the company the IBM World Trade Corporation which later became IBM. He established company songs, a company newsletter, and invested in retaining and training all employees. In short, he changed the CTR culture. In the following years, this resulted in an integrated IBM corporate culture and a "cradle to grave" smorgasbord of educational opportunities for employees. IBM turned out to be one of the most successful corporations the world had ever seen and annually invested heavily in developing and training their customers, employees, and business partners. The investment established a world wide cohesive and strongly integrated corporate culture and mystique. In 1989 marketplace and technology changes signaled that the IBM corporate culture had to be changed, but the key question was how to change it. Some opted for breaking the corporation into smaller business units while others favored keeping the corporation intact. Corporate cultural change started again at the top with a new CEO and new leadership. Long held policies and practices were exchanged for new approaches. Today the IBM corporate culture is not the same as it was years ago, but the IBM Corporation has stayed intact and is thriving. IBM's identity remained intact. IBM divested itself of the IBM Federal Systems Company which was sold to Loral Corporation in 1994. Traditional IBM policies and full employment practices were discarded. The IBM Corporation is still a formidable competitor and has evolved over time to develop an integrated world wide corporate culture. Here are some pros and cons on this approach:

Table 1. Single Corporate Culture

PROS	CONS
One Identity in the Marketplace. An adaptable corporate culture that changes over time.	If the corporation's products become unfashionable or obsolete, the corporate culture can become fractured and unworkable if it's not changed.
All employees and managers know what is expected of them.	Management can become complacent and lack innovative problem solving approaches unless culture is changed.
Corporate teamwork develops with Collaborative behavior. This can lead to less conflict in the workplace.	If culture stays the same, patterns of sameness form leading to inflexibility and resistance to change.

LEAVING CORPORATE CULTURES SEGREGATED

Some corporations tend to leave acquired businesses to their own corporate culture and business processes. They segregate their corporation into a conglomerate of companies. Today many corporations can be defined as conglomerates. Here's my experience with this approach.

Loral

Loral's strategy was to grow through acquisition of other defense corporation's business units. Instead of integrating the acquired corporations into a mainstream Loral Corporation, it allowed each acquired unit to retain its corporate culture. Each retained its own human resource functions, policies and practices. There was much duplication of overhead costs and differing compensation plans, benefits plans, practices, and processes. Loral managed through its strategic planning process and had a very small headquarters staff. Eventually, the corporation divested its acquired assets to other corporations. Lockheed Martin Corporation acquired the remnants of what once were IBM Federal Systems, Goodyear Aerospace, Honeywell Aerospace, Ford Aerospace, Vought, Librascope, Fairchild Aerospace, and others. Later, some of these component companies and business units were divested by Lockheed Martin.

Litton Industries

The Litton Industries strategy was similar to Loral's in that acquired corporations were left largely intact and were not fully integrated into one corporate culture. When Litton acquired Ingles Shipyard business and later when they acquired TASC, the corporate missions, policies, practices, customers, and cultures were so different that they couldn't be easily incorporated into one corporate culture so a practical approach was to provide some guidance and support from the parent corporation when it was required and promote a Litton Leadership orientation for all Litton Industry managers to provide some unifying information about the history of Litton Industries, its major corporate components, and some of its leadership principles. Here are some pros and cons relative to this approach:

Table 2. Segregated Corporate Cultures

PROS	CONS
Segmenting the corporation into former corporate identities keeps the former heritage corporation functioning as it always did. It was business as usual for most of the employees and managers.	Duplicate staff functions led to higher overhead costs, difficulties in transferring between business areas, and lack of a single corporate identity to customers and potential new employees
Acquired corporations with strong customer loyalties maintained their customers	There was often internal competition for markets and customers that resulted in internal battles.

LEAVING SOME ACQUIRED BUSINESS UNITS SEGREGATED AND INTEGRATING OTHERS

Some corporations have adopted an approach to leave some of their acquired business units separate from the main corporation while integrating other acquired business units into the mainstream corporation.

Lockheed Martin

Lockheed Martin Corporation was formed by a merger between Lockheed Aircraft Corporation and Martin Marietta Corporation. Prior to this, Martin Marietta had also grown by acquiring other defense corporations such as General

Electric Aerospace. When Lockheed Martin acquired Loral's Defense Systems business, they initially kept the business unit intact under previous Loral management. Later, part of the Loral defense systems business was divested to form L3 Corporation and the remaining business units were integrated into existing Lockheed Martin business units. This often meant geographic relocation and personnel attrition. Lockheed Martin's long range strategy seemed to be to integrate the differing corporate cultures into the mainstream Lockheed Martin identity. All human resource functions reported to Lockheed Martin Headquarters, and little by little, the differing personnel policies, practices, and procedures were centralized. Lockheed Martin Corporate Headquarters sponsored executive leadership institutes to unify management and frequent corporate wide conferences were held for all corporate training and development components who heard from Lockheed Martin's CEO and executive leaders about the strategies of the company. The internal Lockheed Martin intranet was used to convey leadership development information and the use of the corporate wide training community helped to integrate the disparate corporate cultures. Lockheed Martin's products and services were diverse and grouped into functional business areas such as Astronautics and Aeronautics, Mission Systems, etc. This allowed for some heritage corporate cultures like Lockheed, General Electric, and IBM to remain virtually intact. Here are some pros and cons to this approach.

Table 3. Selective Approach

PROS	CONS
Allows for management flexibility and time to determine whether the acquired business segments should be retained, divested, or re aligned.	Employees uncertain as to future of their career in the company. Multiple HR issues if benefits and compensation system isn't overhauled.
With negotiated agreement, divested business segments can still provide profit for the company	Complex to manage. Staff duplication adds to higher overhead costs, and long term strategies become ambiguous.

INTEGRATING ACQUIRED BUSINESS UNITS INTO ONE CORPORATE CULTURE

Some corporations choose to integrate the people and assets of the acquired business into their mainstream corporate culture. While some best practices of the acquired corporation are retained, the general approach is to suppress the former corporate culture and identity and have employees and managers adopt a new corporate identity and culture.

The Analytic Sciences Corporation

TASC had grown through the acquisition of smaller corporations. Many had not formed strong corporate cultures so integration of them into TASC didn't seem to be a formidable issue. Nevertheless, TASC developed The TASC Institute, a web based corporate institute, that helped integrated the various geographically dispersed components of the corporation throughout the U.S. This strategy helped create a strong cultural identity and strong customer acceptance of the corporation. When TASC was acquired by Northrop Grumman in 1999, the TASC identity was maintained as a sub division of Northrop Grumman.

Veridian Corporation

Veridian also grew by acquiring other smaller defense corporations such as MRJ and Signal. They divested themselves of less profitable business units and developed The Veridian Institute, a web based corporate institute, that was designed to help the CEO integrate the acquired businesses into Veridian through training and development as well as employing various organizational development strategies such as frequent conferences, formation of teams, consulting frequently with senior leaders, etc. When Veridian was acquired by General Dynamics, the Veridian culture was essentially lost and business units of the corporation were merged into the parent General Dynamics Corporation.

Here are some pros and cons to this approach:

Table 4. Integrating Corporate Cultures

PROS	CONS
Corporate Web Based Institute made it easier to integrate acquired businesses into the mainstream corporate culture.	After integrating corporate cultures it becomes more difficult for employees if their business unit is divested and sold.

Table 4. Integrating Corporate Cultures (Continued)

PROS	CONS
One corporate market identity to customers.	When a corporation is acquired by another corporation, it signals the end of that corporate culture and best practices are often ignored.

ANALYSIS OF INTEGRATION STRATEGY OPTIONS

Of all of the corporations I've worked for over the last three decades, IBM had the strongest and most formidable corporate culture. It was able to change with the times and demonstrate flexibility and adaptability. Strong leadership at the top of the organization helped it succeed. Smaller acquired corporations seemed to embrace IBM's approach and few problems resulted in the subsequent integration of those companies into the corporation.

Loral, on the other hand, essentially didn't focus on integrating acquired corporations into a single corporate culture or for that matter, did not invest in human resource development and training as much as other corporations. The result was divestiture of its defense systems business assets to Lockheed Martin, low employee morale, and later spinning off other Loral Corporation business units into separate corporations.

Lockheed Martin initially allowed the corporate cultures of Lockheed, Martin Marietta, General Electric Aerospace, and Loral, to function separately while they learned the best practices of each corporate culture. Later they began the integration of the corporate cultures by combining similar business unit missions, adopting corporate symbolism, slogans, communications newsletters, and establishing central corporate institutes to train all Lockheed Martin executives. Using the Lockheed Martin corporate training community to integrate the corporation seemed to be a successful strategy.

Both TASC and Veridian optimized web based corporate institutes to develop and integrate many different corporate cultures into one corporate identity which seemed to me to be the most successful integration strategy.

LESSONS LEARNED

- Corporations have several choices as to how to organize their business. They can organize by function, by geography, or by products or services. When acquiring other corporations, they must decide on what strategy they wish to employ to either integrate the acquired organization(s) into their existing organization, leave the acquired organization intact with its own culture and organization, or over time, adopt the best practices of the acquired organization(s) and change their corporate culture and way of doing things.

- Plundering an acquired corporation, selling off its assets, and laying off its employees and managers is not an enlightened business strategy. It's an unfortunate carry over from medieval behavior when castles were stormed, inhabitants scattered, and assets looted.

- Other components of integrating multicultural corporations I've experienced included the formation of cross functional teams to analyze and recommend changes in policy, practice, and procedures to top management. Since the process is very complex and often involves technical decisions, customer preference decisions, geographic relocation decisions, and marketing considerations, cross functional teams were often necessary to address all of the components of integrating the newly acquired corporation(s) into the parent company.

- Overall, the main issues in integrating corporate cultures are human resource issues and the best approaches I've experienced involved a focused investment in developing a web based corporate institute chartered in part to help the CEO integrate the corporation and develop human resources.

CHAPTER 3

▼

THE MISSIONS OF A WEB
BASED CORPORATE
INSTITUTE

Traditional corporate departments have a single mission. Web based corporate institutes have multiple missions. The ultimate purpose of the web based corporate institute is to assist the employees and managers of the corporation and enable them to acquire knowledge, skills, and attributes so that they may apply their newly found abilities to help the corporation create wealth for the society it serves and develop a corporate culture which reflects positive human values.

Among the many missions that can be undertaken by a web based corporate institute, continuous learning is fundamental to all of the missions.

THE LEARNING MISSION

An corporation is made up of people and learning can take place at the individual level (i.e. an individual on their own studying and learning), at the group level (i.e. teams of people learning from group discussions), and at the organization level, (i.e. feedback from customers about the corporation's products or services that can teach the corporation how to improve from lessons learned in the marketplace). A web based corporate institute can support learning at all three of these levels.

Knowledge Acquisition

Part of the learning mission is the acquisition, distribution, and application of lessons learned by individuals, teams, and by the corporation as a whole. How does knowledge accumulate in a company, corporation, or organization? Many would answer that it accumulates and is stored in the most talented and experienced people employed in the corporation. Many companies try to store their accumulated knowledge in volumes of documentation, policies, business practices, and standard operating procedures, but for the most part, the documentation becomes obsolete in a short period of time because one characteristic of knowledge is that it changes over time and becomes obsolete and replaced by new knowledge. This cycle time is being shortened by the rapid pace of change in our time. Some companies have databases of who their technical experts are and try to focus on retention of these valuable people assets. New employees often bring in new knowledge and experience and as they gain more knowledge and experience they can and often do take it away with them to their next employer if they decide to leave the corporation. Knowledge can be acquired from experience, research, and lessons learned from past mistakes.

Knowledge Transfer

Knowledge transfer is described as the process by which corporations transfer what they've learned throughout their organization so that they may improve their products and services. They do this by various methods of transferring knowledge and managing the resources and assets of the corporation. Knowledge transfer can take place in meetings, emails, correspondence, bulletins, and other forms of internal communications. Knowledge management can take the form of managing intellectual property. The intellectual property of the organization is often owned by the corporation and restrictions placed on people who come up with new and valuable processes, tools, and intellectual assets that can be profitable for the organization. In recent years, this premise has been challenged by corporate employees who want to benefit personally from their patents, inventions, and original ideas. Some corporations are willing to negotiate and provide shared intellectual property so that the individual may also have ownership of their invention or idea. Many training and development operations engaged in the transfer of knowledge in the corporation do not involve intellectual property of the corporation, but some do, and courses and services become proprietary and not open to people outside the corporation. Some training courses that business partners attend often have non-disclosure agreements signed by the course participants so that course materials of value to the corporation are not disclosed to others. When corporations acquire other companies, the provisions of how intellectual property will be managed and owned are often worked out in advance of the actual settlement.

Knowledge Retention

If employees leave the company, their accumulated knowledge, skills, and experience goes with them. It has become more and more important for corporations to retain highly skilled and knowledgeable people because of the resultant loss in productivity that could result from the loss of key personnel. There is always a discussion that begins with the question of why the corporation should train people who will only leave the corporation someday and all the training given to them and the costs of their training will be lost. *The challenge is not to expect attrition but to retain those talented people.* If they are not trained to do their jobs and grow in their respective fields, they will surely search for a better job in another company that will place a better value on their knowledge, skills, and attributes. *In most cases, people leave an organization because of poor management. The cost of attrition in many cases can be greater than the cost of training that individual.* To retain knowledge in a corporation, management must be trained to understand and respect the importance of each individual person in the corporation. Retaining key people will help the corporation retain the key knowledge, skills, and experience it needs to run the business. A corporation cannot retain key people without giving them the opportunity to develop themselves professionally.

THE ORIENTATION AND TRAINING MISSION

When one corporation acquires another corporation consideration should be given to re-orienting the management of the acquired corporation to the new business environment. New or transferred employees must learn the specialized processes, tools, and methods of the corporation. In short, they must be oriented and trained to do the job the corporation wants them to do with the tools the corporation provides them with. Usually the first encounter a new or transferred employee has is an orientation to corporate policies, processes, benefits, history, and the corporation's job performance expectations. This kind of **informative**

knowledge is meant to help orient the individual to important information about the corporate culture. Most new employees will hope that an experienced person will befriend them and assist them to get started on the job. Some formal training will be required on specialized information technology tools or applications they will need to know. This kind of **formative knowledge** is meant to help change behavior and provide important and useful job skills to enable them to do their jobs. The corporation's values and beliefs should be articulated and understood by all employees. This should include proper professional etiquette and encouragement of civility on the job. Too often there is improper behavior and rudeness that will create a hostile environment and the corporation must help shape the behavior of its employees so as to discourage rude and ignorant behavior in the workplace. The web based corporate institute can sponsor web logs, (blogs) to highlight internal corporate events and news. It can team with the communications department, and play a role in both internal and external communications. Some institutes I'm familiar with that were tasked with the mission of training and development not only addressed all of the current training needs of the company but also addressed the future training needs of the company. At IBM and Lockheed Martin for example, advanced training seminars were conducted with the support of research and development that addressed state of the art future technologies so that a cadre of trained personnel could quickly adapt to anticipated technology innovations. Corporate institutes I've developed had a curriculum for systems engineering, software engineering, project management, technical training, employee orientation, compliance and ethics, marketing and sales, business development, management training for first level, second level and senior management, executive and leadership development. The curriculum was not only adapted to meet current training requirements but also to address trends in those disciplines and future skill needs, as well as incorporating new courses from acquired corporations.

OTHER MULTIPLE MISSION EXAMPLES

A web based corporate institute can be given the mission to address the corporation's research and development needs, act as a clearing house for conference attendance, and/or manage the company's physical or electronic libraries of information. A company having a web-based institute can promote the institute as a repository and distributor of organizational knowledge and find multiple missions for it to perform.

Here are some examples of how different corporations viewed the mission of their corporate institutes.

At Loral Corporation

A center was established at the corporate level. The Systems and Software Resource Center (SSRC) developed performance support databases of company experts in various technical disciplines, technical lessons learned databases for bids and proposals, systems engineering and project management tools databases. For software development organizations, it promoted the achievement of high levels of the Software Engineering Institute's Capability Maturity Model (SEI-CMM). Not only did it sponsor systems engineering and software engineering technical training for the many corporate cultures, but it also provided introductory and advanced project management training and consulting services for all of the corporate business units. This mission was well suited for the Loral environment which did not endorse professional or management development programs. At Loral, the SSRC did not support professional or management development programs which were decentralized in multiple human resources departments throughout the corporation. Since the Loral strategy was not to integrate all units of the corporation into one single Loral Corporate culture, the SSRC crossed over many corporate cultures to fulfill its mission, picked up best practices from each of the Loral business units, and incorporated what was learned into their courses and best practice database.

At Lockheed Martin Corporation

The Corporate Headquarters of Lockheed Martin in Bethesda, Maryland sponsored several corporate level institutes to train executives and managers. At the divisional levels of the corporation, more institutes were formed, many of which had independent web sites and repositories of information for managers and employees. Each division determined the mission of their division level institute. For example, the Skunk Works division formed an institute which focused primarily on technical and business training courses. At the Management and Data Systems division, not only all training and development programs and budgets were centralized at the division headquarters level, but also the division's tuition reimbursement program and high potential technical new hire program was delegated to their director of training and development. E-learning on line courses were provided only to the employees of that division of Lockheed Martin. Management development programs were managed by human resources who received direction and content guidance from Lockheed Martin's corporate headquarters staff. At the division level, management training was sometimes merged with the training and development department and sometimes reported to organization development. Both departments reported to human resources. Each division of Lockheed Martin could acquire other corporations outside Lockheed Martin with Lockheed Martin Corporate concurrence. In most cases, the acquired corporations were small and didn't have strong corporate cultures and so were easily merged into the divisions of the main corporation.

At TASC

New employees were hired and placed initially in The TASC Institute to receive orientation and job training and to await security clearances prior to internal job placement within the organization. All training and development programs including the administration of the tuition reimbursement program and distinguished lecture series were assigned to The TASC Institute. Some research and development tasks were assigned to people awaiting clearances in the corporate institute. New employees received advanced technical training while assigned to the institute. Almost all of the institute's training courses were located on the company's intranet web site allowing all employees nationwide to access training any place and any time. Access to all training materials was through The TASC Institute website which employees could access from the Internet as well as the intranet web pages. The primary mission of The TASC Institute was to provide learning and development and to distribute knowledge throughout the corporation. It also had a customer education mission and sold courses and seats in courses to business partners and customers. Another mission was to reinforce the TASC culture among all acquired and existing business units and to promote leadership development. Leadership development course materials were provided by the parent corporation Litton Industries and modified by TASC who retained its corporate policies, practices, and corporate culture. This approach ensured the TASC reputation with its primary customers and served the parent corporation well. The return on investment was a well trained technical workforce, lower costs compared with the time before the institute was established, customer satisfaction with a highly skilled workforce, and additional revenue generated from the sale of institute courses. Executive and management development was a continuing priority. The TASC Institute was given the mission to track individuals in a succession planning program, create a database to profile each leader, and compile, with the help of executive management, an executive development plan designed and individually tailored to each individual in the succession plan. A web based mentoring program was developed at The TASC Institute that provided workshops, seminars, and matched lists of mentors with employees at all levels of the organization who requested mentoring. TASC also, with the approval of Litton Industries Corporate Headquarters, was able to acquire other companies and integrate them into the TASC corporate culture easily with the support of The TASC Institute.

At Veridian

In cooperation with local universities, the Center for Creative Leadership in North Carolina, and other executive development organizations, The Veridian Institute developed a Leadership Academy. It offered web based tools, workshops, mentoring, coaching, and seminars for employees, managers, and executive leadership. One of its primary functions was to work closely with Corporate Organization Development and other corporate staffs to develop a unified Veridian corporate culture. The Veridian Institute supported the corporation's cultural change initiatives and helped integrate acquired corporations, like Signal, Corporation and MRJ Corporation, and other acquisitions, into a Veridian corporate culture that took advantage of the best practices of those heritage corporations.

Loral, Lockheed Martin, TASC, and Veridian

All of these corporations had similar approaches to pursuing and developing new business. These corporations, like many others in the defense industry, wrote proposals to win new business. Winning proposals have certain characteristics that are worth noting and preserving for future use. Another mission undertaken by a web based corporate institute was to manage data bases of knowledge that helped the corporation win business as well as to improve performance on existing contracts. By capturing the characteristics of winning proposals and successfully managed contracts and making this information available in on line databases and training courses, the corporations were able to continuously improve their performance in these vital areas. Storing this kind of information in databases for future access and use is a job performance support asset. A web based corporate institute must maintain metrics on its activities and employee training records, course materials, tests, and demographic data on who was trained, when, and in what disciplines. This is a requirement when the organization is audited for equal opportunity compliance. Web based training management systems available in the marketplace are an essential asset for the proper management of any training and development organization and assist the corporate institute in maintaining records of the number, type, costs, quality and variety of services performed.

THE MISSION TO SUPPORT EMPLOYEE ADJUSTMENTS TO ACQUISITIONS

Costs, investments, and differences in the ways corporations work are important to understand before an acquisition takes place and after an acquisition is closed. When another corporation is acquired, a total review of the acquired corporation's training capabilities should be undertaken and a major review of its human resource policies, programs and costs should be reviewed. Most corporations will retain the chief financial officer and some of the staff as well as the human resources staff of the acquired corporation until they fully understand the financial operations and human resource differences of the acquired corporation. For example, since tuition reimbursement benefits are often very different from corporation to corporation, a determination must be made of how much the combined corporations will be spending on tuition reimbursement and the differences in tuition reimbursement practices and policies. *My view of tuition reimbursement programs is that they're not always directly linked to the company's skill needs. If it's a benefit, it should be administered as such. If it's part of a corporate institute's mission, it should be considered as a practice, not a benefit, and dedicated to funding only those critical skill degree needs required by the corporation.* Tuition has been dramatically increasing at colleges and universities and corporate budgets for these programs must be increased to keep up with rise in tuition costs. Tuition reimbursement programs are often used by human resources as a recruiting tool to attract new college hires and provide them with financial assistance to continue on for graduate degrees while working for the corporation. *In my experience, they do not help the company retain key personnel, they are expensive, and human resources does not always track drop outs and those employees actually graduating from the college or university under their tuition reimbursement program.* If a corporation needs systems engineers, computer engineers, and software engineers they should reimburse degree costs for those disciplines. Likewise, if they need more MBA's in marketing or finance they should reimburse those degree candidates. Providing a career path for non-degreed personnel by enabling them to attend college and getting reimbursed by the company is also worthwhile if the undergraduate degrees sought will, in fact, enable the individual to move ahead and get promoted into a higher position after they graduate. If an individual does not complete the full program or leaves the corporation voluntarily, a portion of the reimbursement payments should be sought from the individual and repaid to the corporation. Training and development programs must be analyzed. Chief financial officers can collaborate to provide some analysis on the corporation's overhead expenditures and separate these elements from other overhead costs. A web based corporate institute director can recommend the optimum tuition reimbursement plan for the corporation. How much the combined companies are spending on conferences, seminars, and out company training should also be determined. In many cases, the corporation may have separate cost accounts for these activities. When they do not, these elements tend to be grouped into the overall category of training and development which can be misleading. Attending a conference or out company seminar doesn't always lead to building an individuals knowledge, skills, and attributes.

Costs and Return on Investment

Comparisons with other corporations in the same industry should be undertaken. There are ways of comparing how one corporation's investment in training and development compares with a group of corporations in their industry. The American Society for Training and Development's Benchmarking Forum (ASTD), The Society for Human Resources Management (SHRM), and Training Magazine are good resources for keeping up to date on the field of training and development. For most overhead cost expenditures there is little return on investment to the corporation. A corporation must pay its utility bills, facility costs, administrative supply costs, etc. If the corporation counts the overhead spent on training and development as a cost they need to establish how much they want to spend in this area. But most corporations look for a return on their investment in human resource development, which would indicate that they also view this as an investment in their corporation's future. A return on an investment, (ROI), can be calculated not only in terms of how many people were trained in what critical skill areas of key importance to the business, but also in terms of overall return on investment to the corporation in terms of bids won because the people on the bid team were trained to provide an outstanding offer to their customers or commensurately, a reduction in how many bids were lost because the company didn't have the appropriate number of trained and qualified personnel from the customers perspective. Another ROI indicator is increased profitability that directly resulted from the corporation having a trained workforce that improved the quality of the company's products and services. An increase in the number of patents and licensable intellectual properties of the corporation that directly resulted from an investment in training and development is yet another example of how the investment in training and developing people can have a real pay off to the corporation. There are many ways of calculating the ROI of a trained workforce. The consolidation of training and development in the corporation also provides better financial controls and visibility into this important area. In the case of IBM, for example, the corporation initiated a corporate wide investigation to find out how much they were spending in this area which resulted in improved management and economic benefits to the corporation. In the case of TASC and Veridian, the development of web based institutes resulted not only in cost avoidance but also direct savings on overhead compared to the time before the institutes were developed. Some corporations have chosen not to have a budget for training and development and have "cost centered" this activity. Cost centering is an accounting approach that basically works as follows.

- Staff labor costs, space, educational resources, or materials devoted to training and development and other institute services are not budgeted at the department level.

- Costs for training and development and the institute are pooled by accounting in a centralized cost center or account. Every month, costs of the institute are accumulated in this account.

- Costs of the institute are recovered by the institute staff essentially selling its services to the rest of the corporation on a "pay as you get the service" basis. When an organization sources services from the institute, they spend their overhead dollars into an internal corporate training "receivables" account in the corporate institute. The goal is to recover the costs of the institute from payments received from the rest of the corporation.

- It helps if the overhead rates for training and development are lower than the overhead rates for the rest of the company and sometimes this is possible. The suggestion to centralize the institute at the highest level of the corporation and spread its costs over the entire overhead structure of the company can put the institute into a general and administrative, G&A) category rather than an overhead category. For example, one dollar of engineering overhead costs one dollar to the engineering function. If the cost of the institute is spread over three functions let's say, engineering, manufacturing, and finance, then the engineering function pays 1/3 of the dollar of overhead allocated to source its training requirements from the institute. While the total cost picture all comes together at the corporate level, the perception at the engineering functional level is that they're paying less of their overhead budget for institute services rather than to pay out a larger amount from their own overhead budgets. Federal Acquisition Regulations will stipulate and define allowable and unallowable costs and define the different cost categories.

- If the training staff is successful it will recover all of its costs during a calendar year. If it appears that they will not recover their costs, the institute must reduce its operating costs to equal what they recover from the corporation. If they recover more than their operating costs, they might be able to expand their services the next calendar year.

Some cost centered ventures work, but it requires a lot of discipline, cost accounting, lower negotiated overhead rates, and a dedicated staff. In the corporation where I've seen it work, travel budgets tended to go up, cost accounting was a consuming focus, and development of new services tended to decline. *I would not recommend this approach*. If a corporation is committed to establishing a centralized corporate institute, it should provide the institute with an annual budget that is prioritized and addresses the needs of the corporation. Too often, some of the corporation's departments that have budgets to procure training and development services have higher overhead budgets and those departments without overhead budgets cannot procure training and development services for their staff. This prompts inter organizational conflicts and issues.

Retention of Human Capital

Retaining valued human resources is always a priority. A corporate institute also helps to attract and retain new talented people. In my experience, the development of web based corporate institutes was highlighted in the corporations recruiting literature and resulted in increased interest on the part of new hires who were applying for jobs with the corporations and who favored continuing their education. Although there was little information to suggest that the formation of the corporate based institutes were instrumental in retaining employees, there was sufficient data to indicate that many leave corporations when there is a lack of formal training and development opportunities so I would conclude that they help retain employees.

THE MISSIONS OF TRAINING DEPARTMENTS

Training departments won't enable the successful integration of corporate cultures in the corporation. They are very different from corporate institutes and are rarely in tune with the mainstream business of the company. While training and development is an important if not key feature to the successful integration of multicultural corporations, training departments cannot function with the same proficiency as a web based corporate institute. In my experience at IBM, I managed training departments which were historically characterized by the following activities:

- The introduction of computer assisted instruction, (CAI) in the 1960's. CAI offered IBM a way of providing tests and examinations as well as remedial training in basic topics like math, reading, and writing. At some locations, pre and post tests were administered by CAI and remedial training in reading and writing was provided to employees seeking high school equivalency training.

- The formation of learning resource centers took place in the 1970's. At several IBM locations learning resource centers were attached to site libraries and administered by the location's training department. The learning centers contained self study materials, audio and visual cassette materials, and access to CAI courses as well as videotape courses.

- The introduction of personal computers, (PC's), satellite, and interactive broadcast television, (ITV), emerged in the 1980's. With the introduction of personal computers, training departments set up PC classrooms to train their own employees on how to use them. The technology was new at the time and in addition to managing the traditional professional development, voluntary education, tuition reimbursement, and skills training programs, PC training was added to the mission of the training department.

- Interactive broadcast television as well as satellite delivered education also came into use in the 1980's. Some IBM facilities had interactive satellite education programs which focused primarily on marketing education. At

IBM plant and lab locations, IBM training departments worked in collaboration with colleges and universities who were experimenting with broadcasting interactive television courses from the campus to the IBM facilities. In the Washington D.C. area, universities like The University of Maryland and The George Washington University developed these capabilities. Later other colleges and universities in Virginia such as Virginia Polytechnic and State University and the University of Virginia also developed this capability.

- The Internet and networking technologies introduced in the 1990's added to the technical training mission. Programmer training requirements of the previous decade were supplanted by systems engineering and software engineering programs and there was accelerated interest in leadership development and training project managers. By this time distance learning was developing as another mission of the training department and new companies that specialized in developing "courseware" for distance learning were coming into being.

From the mid 1980's through the 1990's corporations started to establish corporate universities and started to centralize and focus their training and development investments on aligning that effort with their corporate strategic initiatives. It was at this time of intense and accelerated technology development that I was fortunate to direct the training initiatives of a major division of Lockheed Martin, and later develop a corporate institute for TASC that was partially web based and a corporate institute for Veridian that was largely web based.

Characteristics of Training Departments

I was fortunate to be able to manage training departments as well as institutes in my career. The characteristics of training departments were as follows:

- *They were largely administrative functions* and the mission was mostly to coordinate training requests from management.

- *They were largely decentralized, tactical* and managed the priority of the moment, whether that happened to be a contract with the National Alliance of Businessmen to train disadvantaged, unemployed people, or to manage a programmer retraining initiative to cross-train administrative employees to become programmers. Other functional organizations in the organization had their own training and development budgets.

- *They were human resources departments* where professional HR personnel rotated into and out of on their way to becoming human resource managers. Training and development was not seen to be a career unto itself and career opportunities in the training and development profession were highly limited.

- *They were classroom-focused* where scheduling and coordinating courses were offered to a wide audience when the training department was ready to offer them, and,

- *They were not always outcome or quality focused.* There was little to no benchmarking accomplished. Management in human resources was not particularly interested in the quality of the training program unless there were complaints from management or employees about the course(s), and little to no data was collected regarding the training program outside of counting how many student days of training were offered and how much it cost.

- *Very few presentations were given to top management.* Human resources provided most of the presentations to top management and training and development were often added as an afterthought.

THE MISSIONS OF CORPORATE INSTITUTES

Web based corporate institutes can take on many roles and missions. Some may want the web based corporate institute to only offer E-learning courses in information technology areas, particularly those involving certification in an area of high interest to the corporation. Others may want the web based corporate institute to accomplish a host of tasks and missions.

Characteristics of Web Based Corporate Institutes

The institutes that I've developed and managed tended to be:

- *A cultural change leader* in the organization. Corporate executives took a strong interest in developing a corporate institute that would optimize technology, reduce costs, and help change the culture of the organization. An emphasis on leadership development was seen as an enabling initiative to change the corporation's culture along with a strong communications program and interventions by the company president or CEO. They were less administrative and more technically driven.

- *They were centralized and strategically focused* at the highest levels of the organization. All budgets were centralized in the corporate institutes and focused on strategic priorities established by top management.

- *They were strategic* and tied closely to the corporation's strategic direction. The institute director was part of the top executive leadership team and sponsored careers in the training and development profession. *Both institutes I developed reported one level below the company President and one level below the company CEO respectively.*

- *They relied on distance learning technology and were not primarily focused on classroom delivery.* The staff of the institute included developers, instructors, and deliverers of creative learning solutions that optimized learning technologies.

- *The web based institute focused on specific knowledge, skills, and attributes to be learned* and were driven by a desire to measure outcomes and assess the quality of their programs. There was a focus on process improvement and quality.

- *They benchmarked, compared, and evaluated* their programs on a regular basis. Metrics were collected and presentations were made by the institute director to executive management on a regular and frequent basis.

Unlike training departments that for the most part report to HR, web based corporate institutes in most cases do not report to human resources organizations. All other human resource training and development programs and all decentralized budgets are centralized and consolidated in a corporate institute.

RATIONALE FOR THE MULTIPLE MISSIONS OF A CORPORATE INSTITUTE

There are many missions that a corporate institute can accomplish other than training and development. With the right leadership and proper resources, successful accomplishment of all assigned missions will lead to a return on investment for the corporation. An institute can provide access to new learning technologies that promote performance based learning and lessen the need for costly classroom training. A web based corporate institute is connected to all of the mainstream operations and support functions of the corporation. An institute can be not only an investment made by the corporation for human resource development, but also an asset to winning contracts, training customers, and developing a corporate culture. A web based corporate institute can help change the culture of the corporation. In fact, it is, in my opinion, the reason why some corporations are more successful at integrating their corporations and others are not. The primary mission of a corporate institute is often training and development but web based institutes can be focused on research and development, process improvement, and cultural change. Missions of corporate web based institutes can be expanded to include succession planning, organization development, and administration of company education policies and practices. Increasingly, institutes have their own intranet and Internet websites and are finding new ways of providing learning options to employees, managers, and customers.

Without a clear rationale and vision that explains how the corporation can benefit from establishing a web based corporate institute the initiative will not be readily accepted by corporate management. The rationale should reflect what the corporation believes and values in terms of human resource development. The corporate institutes I'm familiar with have a policy statement or charter on the web site that articulates the corporation's rationale for the web based corporate institute. *The formation of a corporate institute is a long term commitment. An important message is sent to the customers, employees, stockholders, and managers of the corporation when an institute is created. That message is that the corporation has a long term investment in its employees, managers, customers and stockholders. The corporation wants all of its constituents to recognize the uniqueness of their corporate culture and the value the corporation places on the human side of the enterprise.* After formation of the corporate institute it would be a mistake for the corporation to put itself in a position to have to eliminate the institute and retract its commitment at a later date due to lack of overhead funding for the institute.

The corporation needs a corporate institute to accomplish the following:

- *Attract and retain key personnel.* New applicants are drawn to corporations not only because of increases in salary or benefits, but also are keenly interested in the kind of corporation they are applying to and how it expresses interest in their development and well being.

- A web based corporate institute helps *integrate newly acquired corporation personnel into the parent corporation and develop and change the corporate culture* as needed.

- *It consolidates all training and development resources and funds in one place.* It is remarkable to me that many companies have permitted the decentralization and liberal use of overhead funding in the development of human resources without considering this as a strategic investment. Research and development, (R&D), funding, for example is usually centralized at the top of the organization under a Vice President of R&D and departments in the company must justify their R&D budget requests and obtain funding for their projects from a centralized source. This process should also be the way to manage a web based corporate institute.

- *A Web Based Corporate Institute provides a "one-stop shopping" source for employees to find training and development opportunities.* Employees often do not have time to search around the corporation for funding or programs to solve their knowledge and skill needs. Turning to a centralized, on line, web based corporate institute solves this problem for them and saves them time. Getting web based information on the progress made in integrating a newly acquired corporation into the parent corporation is also an asset to help integrate multicultural corporations.

- *A Web Based Corporate Institute provides all employees with a belief that the corporation encourages their continuous acquisition of skills and knowledge.* Many employees in many corporations today do not believe that their company encourages their professional development.

- *It provides employees with the opportunity to cross train into other careers.* Training for employability is important in an age where full employment cannot be guaranteed by the corporation.

- *It provides a forum for internal and external executive speakers to offer their guidance and perspective to employees on company values, ethics, and problematic issues.* Executives need to reach, teach, and preach to their employees about the corporation's strategic plans, opportunities, and the future. A web based corporate institute can provide web casts where executives can reach the entire employee population or a segment of it.

- *It provides for knowledge to flow from outside the organization into the organization.* A web based corporate institute will have multiple supplier's business partners, and relationships with universities. Using the Internet to gain world wide access to knowledge can be a valuable asset to a corporation.

- **It provides a forum to promote the cultural values of the corporation and provide continuous, life long learning opportunities.** It goes without saying that learning is not a one time event and that a continuous flow of knowledge is needed by most professionals in corporations.

- *It can help management segregate overhead costs of doing business into "investment-like costs" versus "maintenance-like" costs so as to better understand and balance how overhead is spent.*

- *The Web Based Corporate Institute offers a clearinghouse for scientists and engineers to collaborate and share knowledge and research findings.*

The multiple missions which can be assigned to a corporate institute can evolve over time and opportunities to link up with other public institutes, universities, web-based resources, and corporate partners can broaden the scope of the institute. Partnerships formed with specialized training vendors and universities will strengthen the quality of the training and development programs offered by the company. Each corporation will, in all likelihood, have a different type of corporate institute depending on its culture. The platform for providing corporate institute services and products is the company's intranet.

In short, the institute should focus on the strategic needs of the company. As management and employees experience the success of the institute, they will adopt it as their own. Employees and managers involved in contract bids can illustrate the company's commitment to training by including the institute in its proposals. The institute can become a profit center and customers can be offered courses and consulting services from the institute.

LESSONS LEARNED

- Different corporations have different approaches with regards to training and developing their most important asset, their employees. Consequently there are different missions and priorities that corporations can delegate to a web based corporate institute. Cost centered approaches generally don't work.

- Training departments are very different from corporate institutes and generally are too narrowly focused and cannot provide solutions to help integrate the corporation following a business merger or acquisition.

▼

CHAPTER 4

CHALLENGES MET BY CORPORATE INSTITUTES IN DEFENSE CORPORATIONS

Speed, flexibility, and adaptability are important to defense industry corporations that operate in an intensely competitive environment. Overhead costs must be reduced but valued overhead investments like human resource training and development retained. A corporation must be willing to adopt the best practices of acquired corporations and be ready to make major changes to its corporate culture. Some corporations are forming web based corporate institutes to help address these challenges.

Mergers, divestitures, and acquisitions have been rampant in the defense industry from the 1990's and early part of the 21st century to the present. *Many defense corporations have found that the acquisitions are unfinished. They are finding they have to manage separate corporate cultures, different ways of doing business, duplicate staffs, conflicting organization policies, and they're facing a variety of internal problems.* The corporations had to decide whether to leave each acquired business unit operate as a separate business within the corporation, integrate it into one business, or plunder it's assets and layoff people to cut costs.

Today, corporations must be able to quickly change directions to adapt to market conditions and even change their corporate culture to remain competitive. I'd like to share some examples of divestitures, mergers, and acquisitions from my experience and how some corporations have formed web based corporate institutes to help them find a solution to the challenges they face following corporate mergers and acquisitions.

IBM CHALLENGES AND APPROACHES

IBM in the 1980's was at its peak. Admired by the business community for its products, integrity, and financial performance, ranked by college professionals as the number one company they hoped to work for, employing roughly 400,000 people, and generating about $50 billion dollars a year in revenue, it was the corporation that everyone admired. Three simple beliefs guided the corporation and were communicated throughout the company's literature. In brief they were:

- *Respect for the Individual* is a belief that people, not necessarily money or material things, is a corporations greatest asset and people should treat each other respectfully.

- *Providing the Best Customer Service* was everyone's job. Assuring that every employee's job related to enhancing the goal of customer service would help secure the corporation's reputation and ensure full employment, and

- *The Pursuit of Excellence* is a belief that superior performance was required and expected of all employees in both product and service areas.

At that time, IBM conducted customer training at over 100 major education centers in the world. An interactive satellite system, the Interactive Satellite Education Network (ISEN), connected major training centers throughout the United States and the world. In Europe and Asia, hundreds of thousands of student days of education were delivered annually at approximately 18 major education centers in 11 Asian and Southern Pacific area countries. On any given day in the United States, 18,000 of the companies 390,000 employees were enrolled in a formal training course. In 1989, about 5 million student days of training were conducted which averaged 12 days of training for each IBM employee per year. IBM Corporate Headquarters located in Armonk, New York, was the location of two major training centers. The IBM Management Development Center and the

IBM Thornwood Education Center sponsored training courses for the entire corporation. The Thornwood Center contained a library, video and satellite transmission studios, a gym, recreational areas, and housed student employees overnight.

IBM went through a major and difficult financial and organizational transition in the early 1990's and was almost split into smaller companies. In 1993, more than $8 billion dollars of cost reduction were undertaken. But the talented people of the IBM Corporation under the leadership of Lou Gerstner enabled the company to continue to exist as one entity and bounce back from the brink of bankruptcy to become a profitable company today. The scope of education and training across all IBM divisions world wide, the synergy between the many education centers at manufacturing plants, labs, and sales locations, created a cultural environment of continuous learning that served the company well in its come back from the brink. The defense systems component of the corporation was the Federal Systems Division, later called the Federal Systems Company, (FSC). With cost cutting measures introduced in the early 1990's, the Federal Systems Company, which had about 11,000 employees, was sold to Loral Corporation. The divestiture of FSC to Loral had profound cultural implications on the IBM population. It signaled the end of the IBM culture that never had layoffs or sold major divisions in its entire corporate history up to that time. Many people left the Federal Systems Company after the divestiture. Some tried to stay in IBM. Others had to learn to adapt to the Loral corporate culture.

LORAL CHALLENGES AND APPROACHES

Loral opted to have a decentralized approach to managing the company and chose to keep acquired business units of companies relatively intact with some management process changes. Loral's diverse division missions, products, and services were applied to support defense and government programs. After the IBM Federal Systems Company acquisition, defense contracts were merged with the Loral divisions working contracts in the same areas, but for the most part, the IBM FSC business unit of Loral was left intact. One of the overhead costs targeted for reduction in the acquisition was the Software and Systems Resource Center, (SSRC). This center revised their original mission to support all of the acquired and existing Loral divisions involved in software development. The SSRC would provide expertise and assistance in improving processes, tools, and program performance. Some of the acquired companies included business units of Goodyear Aerospace, Ford Aerospace, Fairchild, Vought, and Honeywell, among others. Loral's challenge was to manage all of these different company cultures, business processes, and HR practices. The primary objective of the Center was to prove to management that program performance, new contract win rates, and improvement in the company's competitive position in the marketplace could be directly linked to the successful implementation of the SSRC training and consulting mission. The center operated as a corporate institute and provided an information repository accessible to all Loral divisions which contained common processes, standards, documents, training, and software tools and information. The intranet based repository also provided links to other relevant servers on the Internet such as the Software Engineering Institute and the Defense Information Systems Agency. The Center's staff that was comprised of highly skilled, multi-disciplined professionals from the different Loral divisions and cultures, analyzed processes, methodologies, and tools and recommended best practices to improve software and systems engineering operations. The Center established technical teams that provided specialized technical training and development in the disciplines of systems and software engineering and program management. The Center maintained a database of corporate technical experts, and provided consultation and support for new bids and existing contracts. The SSRC was able to show a substantial return on investment to the corporation after one year of operation. Several business units achieved high level ratings from the Software Engineering Institute enabling them to bid and win software development contracts with the Federal Government. Since the SSRC was a cost center, it had no budget and recovered all of its overhead costs from internal clients so there was no corporate funding for the Centers products and services. The role and mission of the SSRC fit into Loral's strategic plan. It reported to the Vice President of Technology at the corporate level and helped integrate best practice solutions throughout the corporation. The web-based strategy of the Center, however, did not include E-learning nor could it take advantage of web technologies which weren't available until the late 1990's and early part of the 21st century. While the cost-centered approach to funding the operation worked for two years, it resulted in less of a focus on the future and less investment in updating course materials and developing new materials.

In most cases corporations have no single function or organization in place to address the integration of the business units. Cross-functional teams and committees are formed to make recommendations to executive management on what should be done. Most of the time, the integration of other business units into the mainstream culture is poorly done and costs the corporation in terms of increased overhead costs and attrition of key people. Conflicting corporate policies in many cases can lead to confusion. A web based corporate institute can help by providing information sharing and cross organizational communications and guidance. In Loral, the process of identifying goals common to all divisions and have the SSRC function as the coordinator of cross-functional and cross-organizational issues as well as to address major technical business needs was a practical way to integrate the corporation but it was a carry over from the IBM corporate culture and not viewed by Loral as an integrating function. Business units still retained their autonomy. Overhead costs were reduced overall through compliance with a centralized strategic plan. After two years of operation in Loral the SSRC started to lose its value to the corporation. Two years after the IBM acquisition Loral divested the defense systems sector and the SSRC to Lockheed Martin. Employees, all acquired by Loral from different corporate cultures, were forced to adapt to a new corporate culture and changed corporate identities once again.

LOCKHEED MARTIN CORPORATION CHALLENGES AND APPROACHES

Lockheed Martin Corporation is a highly diversified global enterprise engaged in the conception, research, design, development, manufacture, and integration of advanced technology products and services. Most of its customers are departments and agencies of the U.S. Government and it is widely recognized as a premier defense contractor. Lockheed Martin had a centralized approach to managing the corporation but it strategically allowed some operational units to maintain their former cultural heritages and gradually identify with Lockheed Martin. For example, the Loral SSRC continued to operate as usual and expanded its services but about a year later the SSRC was disbanded and scattered throughout the Lockheed Martin major business divisions. Most of the SSRC training curriculum was acquired by the Management and Data Systems, (M&DS), business division of Lockheed Martin. This business unit was formerly General Electric Aerospace and had its own business culture and operations. M&DS modified the SSRC courses and established a budget to convert the training materials to the needs of M&DS which retained many of the characteristics of General Electric Aerospace. Some of the heritage General Electric training programs were unique to this division at Lockheed Martin.

THE ANALYTICAL SCIENCES CORPORATION CHALLENGES AND APPROACHES

The Analytical Sciences Corporation (TASC, Inc) was a division of Litton Industries when I joined them in early 1999 and was recognized for providing advanced information technology solutions for government and commercial clients in areas such as network and infrastructure security, systems engineering and modeling and simulation. The TASC Institute for Learning and Development, abbreviated in the literature as The TASC Institute, had a training center and was also a web based institute in that most of the training and development courses were provided over the corporation's intranet. The TASC Institute used computer-based training, video-conferencing, various distance learning technologies, and traditional classroom training. Since TASC was recognized as one of the worlds premier systems engineering and systems integration corporations The TASC Institute sponsored a Certified Systems Engineering Program, on line technical certification programs, a software engineering program focused on new software technologies, processes, tools, and methodologies, and a distinguished technical lecture series featuring noted technical experts from academia, national laboratories, industry and government. The TASC Institute reported to the Vice President of Administrative Services, (one level below the company President). Top executive management participated in The TASC Institute programs and called the largely web based institute a best practice in the Litton Industries Corporation. The institutes classrooms and labs were located adjacent to research and development labs. Promising new technical leaders were hired and placed directly into the institute. They were provided with advanced technical training, and later placed in key positions in the corporation. The program was named the Technology Leadership Development Program and was widely praised by TASC management. The institute director served on research and development and systems engineering committees and was directly involved in the corporation's mission. In my opinion, reporting the institute to the Vice President of Administration who was also responsible for facilities planning and human resources was also a mistake in that there was a closer relationship between the institute, research and development, the operating business units, and engineering than between the institute and facilities and HR.

VERIDIAN CHALLENGES AND APPROACHES

Prior to being acquired by General Dynamics in 2003,Veridian was a leading provider of information based systems, integrated solutions, and services specializing in mission-critical national security programs for the intelligence community, the Department of Defense, law enforcement, and other U.S. government agencies. The Veridian Institute was developed primarily as a cultural change agent and provider of innovative, high quality learning solutions in the areas of business development, project management, technical training, compliance, and leadership development. The Vice President of Human resources and the Director of The Veridian Institute reported to the Vice President of Organization Development who in turn reported to the CEO. This created an important link and synergy with the corporate staff but reporting to a Senior Vice President of Organization Development was awkward in that organization development was viewed as a corporate initiative and was not as directly connected to the operating business units as the institute. Organization development typically reported to HR in most corporations but because the institute was formed to integrate the corporation into one culture, reporting the institute and HR to Organization Development served to strategically focus its mission and provided direct access to the CEO.

From its inception, The Veridian Institute was viewed as a web based institute and 80–90% of its products and services were web based. The institute was provided with an external website and customers could view the products and services offered by the institute on the Internet. Customer training courses were planned for web-based access and sales and the institute was referenced in bids and proposals to assist in obtaining new business. The institute also had a robust intranet website accessible by all employees and managers. The intranet website provided a catalog of all training courses offered, hundreds of E-learning information technology courses, on line compliance and ethics courses, registration procedures, policies, practices, and links to access other performance support information. Also on the web page of The Veridian Institute, the CEO of Veridian provided a video web cast to all employees supporting the institute with his personal endorsement of its products and services. The web based institute fit very well into the corporate strategy and received strong support from senior management in the corporation.

LESSONS LEARNED

- Low cost, speed, and agility requirements of today's corporations require the formation of an institute that is flexible, optimizes the use of distance learning, and is directly aligned with the company's mission and goals.

- Forward looking corporations will not report a web based corporate institute to human resources or other staff functions that aren't directly connected to the strategic and operational elements of the business.

- Examples of corporate education as I've experienced it in five corporations, IBM, Loral, Lockheed Martin, TASC, and Veridian show that there is much diversity in mission assignments and reporting relationships. While each has undertaken a different path to human resource development and each has strengths in its approach, the optimum approach would be to report the institute one level below the highest level in the company preferably to an operational executive rather than a staff executive.

- The institute should reflect the corporation's culture as well as help shape it.

- The technology of the time and the robustness of the Internet determine the capabilities of the corporation to "webify" most of the training and development support activities assigned to the corporate institutes. Most of the corporate institutes today, are not completely web-based institutes. Most have a combination of web-based and classroom based services.

▼

CHAPTER 5

HOW TO DEVELOP A WEB BASED CORPORATE INSTITUTE

> **The process to develop a web based corporate institute closely follows the systems engineering process which begins with establishing a baseline and identifying requirements and follows with the phases of design, development, testing, logistics support, and continuous process and quality improvement.**

Web based corporate institutes represent an innovative approach to the future of human resource development, technical training and customer education in corporations. However, if they're not developed and supported systematically, they are not very durable and often like sand castles are washed away with a tide of new management. To be successful, the web based institute must be developed with systems engineering processes and methodologies familiar to systems integrators. The first step in developing a web based corporate institute is often the most difficult since it involves convincing management that it is needed. To develop a web based corporate institute, this systematic approach is recommended.

ESTABLISH A BASELINE

Executive management must be committed to the formation and development of a corporate institute. The institute cannot be the brainchild of the human resources department or the engineering function, the marketing manager, or any functional organization in the company. The CEO of the company must be committed and willing to commit his or her own personal time to its success. There will be critics of a corporate institute and those that will want to undermine its creation. Only the CEO's backing will sustain the early formation and longevity of the institute. In the event of a merger or if the corporation is acquired by another company, the CEO should take the lead to assure that the corporate institute will transform itself to support the requirements of the new parent corporation.

The first step I've always taken when developing a corporate institute is to assess what the corporation is currently doing to train and develop the workforce and what it is spending on this effort. Management meetings with the leaders of human resources, finance, information systems, accounting, and the operational divisions of the corporation follow and a survey undertaken to assess what they believe the corporation should be doing to improve overall business performance. If mandated by the CEO that the corporation will have a corporate institute, enlisting executives to serve on the institute's advisory board is one way to win support for the institute. Most of the time an executive advisory board will not agree on the strategy or plans for the institute in which case the CEO must act as the tie-breaker as serve as the board chairman.

The accounting manager or controller may argue that forming a corporate institute will constitute an overhead burden. The first step in getting the CEO's support, therefore, is to understand the baseline of how much it currently costs the company to train and develop its workforce and show how consolidation of training and development budgets in a web based corporate institute will lower overhead costs. How are the corporation's overhead funds allocated? If they are spread throughout the organization, accounting should identify what is budgeted for training and development and where the budgets have been allocated. Are the budgets allocated equitably? Too often, there are some organizations in the company who get no allocated budgets for training and others who get more than they need. Allocating budgets to first level managers is an error that most company's make in the mistaken belief that the first level managers know more than the senior level managers who needs what training. Many employees may tell you that their managers have no idea what their individual knowledge and skill needs are. *This approach addresses short term tactical training and development needs but since first level managers do not see the big picture and have no overall view of the strategic needs of the corporation, they're not in a position to determine the overall strategic training needs of the corporation.* How is the money spent? It may come as a surprise to find out that the corporation has multiple contracts with the same training vendors and is spending more than what they should for the same education and training services. This is particularly true if overhead budgets for training are allocated to the lowest levels of the organization. Duplicated courses, duplicated vendor contracts, even duplicated and expensive E-learning contracts have been exposed in many corporations who have never done this kind of assessment. Finance may become the easiest organization to convince that a consolidated budget for overhead training and development expense is better for the corporation to manage and control and it's always a good idea to include the top financial executive on the advisory board of the institute.

Human resources may argue that the institute is not needed and that the human resources traditional training department that now exists, if one exists, is sufficient. This same argument might also be brought forth by marketing for sales training, by procurement for their personnel, by manufacturing, and by the engineering function. This position argues that a decentralized approach is less costly and more specifically oriented to solving functional training problems. *This position, too, is tactical, reactive, misguided, and under close scrutiny reveals that most human resource "training" and "management development" programs are orientation and indoctrination sessions which do not build the leadership skills or management skills needed by the company, and most functional training programs duplicate costs and courses.* They operate under a presumption that their "towered" organization needs to only address short term needs. While most staff departments are cross functional they do not have the training and development expertise needed in a modern corporation. Some might argue that it doesn't matter where the institute reports in the organization but I would disagree. It matters a great deal and must report to or one level below the CEO in order to have a significant impact on the corporation.

Many corporations are trying to build teamwork in a corporate organization that is largely vertically structured. They are trying to minimize the bureaucratic influence of these heritage vertical functional towers. Cross functional teams and information sharing is vitally important in corporations that are characterized by rapidly changing technology and marketplace conditions. *In order to be effective corporations must operate as a total enterprise not as isolated and often warring factions. Centralizing overhead costs is one way to encourage interdependence on the organization and interdependence fosters cross functional teamwork*. Knowledge and information must be shared throughout the corporation and not captured and hoarded in a single staff function.

Once a baseline has been established and initial meetings with all key executives conducted, a presentation should be given to the CEO and later to all top managers of the corporation that outlines the findings of the survey, analysis of the corporation's issues, the vision of the corporate institute, and how competitors are addressing the same issues. If successful, all of the management team should begin to support the development of the web based corporate institute.

DETERMINE REQUIREMENTS

In this stage of development, three major items need to be addressed; assessment of the corporation's training and development requirements, an analysis of the requirements, and determination of resource requirements. Too often, corporations have never done a thorough job at determining the training and development requirements for the people in their corporation or determining what a corporate institute could be tasked to do for the company. Often, no annual needs assessment process has been implemented and no skills, knowledge, and attribute definition been made for each job. Job descriptions, in many cases, are so vague that it's often difficult for recruiters to find the right people for the job openings in the corporation. An engineer, for example, may require different skill sets depending on the job. Information technology skills may be required by almost everyone in the company, but not all employees require skills in the same software tools and applications. A training needs assessment survey should capture the skills, knowledge, and attributes employees require to do their jobs today and those they will need in the future. This process should begin as a dialogue between the manager and individual employees that report to them. They should then develop an individual education or training plan. When done correctly and summarized, the training needs assessment survey will capture all of the department education plans and reveal all of the corporation's skills needs and training requirements. Commercial information technology applications and tools are available to assist in this effort and sophisticated training management systems are on the market. They often have a training survey that lists all of the currently available courses a corporation has in its inventory, a template for an individual education plan, a skills inventory, and the capability to distribute it to all managers over the intranet. Managers and employees can check off those courses that are high, medium, or low priority needs. If a course doesn't appear on the inventory list, new courses required can be added. Employees can assess their own skill gaps and select training that will help close their skills gaps. This web based bottoms-up approach is matched with a top-down approach whereby top executives are queried and surveyed to determine what overall leadership training is required. Their views on why important bids were lost because of customer perceived skill deficiencies and what skills are needed that the corporation doesn't possess will help define the corporate priorities. An analysis must then be undertaken to determine the major training and development priorities that will be addressed in the current budget year. *Not all of the training and development requirements can be met in a single budgetary year and training is not the solution to every skills gap problem so the assessment needs must be prioritized and those that are most closely aligned with the success of the business need to be addressed as a priority. Skill gaps can be closed by other means such as staffing, temporary subcontractors, mentoring, or by providing on the job training.* Analysis will also help determine when the needed training is really required (i.e. immediately, first quarter, second quarter, etc.). Budgets can then be prioritized for a particular year and the corporation will know who is being trained on what, when, and where the money will be spent. Since everyone will not be trained or developed in a single calendar year, many corporations aim to train at least 50% of their population every year. *The corporation should make it known that it is the employee's responsibility to obtain the training and development that they need whenever they need it to keep current in their occupational skills.* A good training management system will maintain records to summarize the results of employee initiatives to close their skill gaps. Completions should be noted, tracked, summarized, and reported quarterly to top management. Both formative and. informative training should be part of an employee's education record. For example, new employee orientation normally required for all new hires and transfers into the corporation and new manager orientation to the corporation's policies should be recorded. Formative training directed at skill building and behavior change such as project management training, business development training, information technology training, software tools and processes, and systems engineering education along with safety and health training should be part of an employee's education record. Specific training on new systems is required whenever new systems are being installed or new technologies introduced. E-learning sponsored by the web based institute and available on the corporation's intranet is one way to provide on going information technology training. An analysis of how the required training was delivered and the type of training provided should also be maintained and reported.

Determination of resource requirements needs to be undertaken. Vendor contracts, instructors, materials, and budget requirements need to be understood and developed. Instructors must be qualified to teach. ***Too often, corporations believe that telling and teaching are the same and they are not. Not everyone has the training and qualifications to be an instructor.*** A web based corporate institute can determine the qualifications of instructors and train them. Economical use of instructor resources must be determined so that course costs don't escalate and drive overhead costs higher. Distance learning technologies have improved in the last decade to the point that many employees would rather participate in web based learning rather than have to leave their work locations and travel to a classroom. Using web cameras (webcams) instructors can deliver seminars over the web from their own offices or workstations. Installation of appropriate software at employee work stations can provide audiovisual contact with the instructor. "NetMeeting" and other software applications can provide visual and audio support through most computers and there are videoconferencing systems that are useful as well. In addition, distance learning simulcasts can be set up so that all participants can see and hear the entire session. Instruction can also be provided in web logs, (blogs), which are bulletin boards set up over the Internet or intranet to provide specific information. Knowledge can be shared quickly and efficiently by trained and qualified instructors with employees geographically located far apart. Often employees learn from each other in these sessions.

Not only do training and development requirements have to be determined, but also the instructor, materials, labor, and travel requirements need to be assessed so that accurate budgets can be developed for a particular year. Internal versus external instructor costs will vary and institute professionals can determine the market value of instruction in specific areas. In general, managers and executives should not be compensated for their time in a classroom or in a web cast as instructors. For specific, high-level technical courses, employee instructor compensation may be warranted, particularly if the course requires preparation and conduct beyond the current job commitment of the individual. For specific corporation courses that need to be converted to web based delivery, the institute staff can work closely with the corporation's web master to "webify" the course. All course materials should be the property of the corporation, not the instructor since the company is paying the instructor a salary and paying for the time and materials to develop the course. Fair Labor Practice laws and standards, corporate policies, government standards like the Federal Acquisition Regulations (FARS) that determine allowable costs for training, and other specialized experience in addition to adult education and teaching experience is required by the corporate institute staff. The corporate institute staff must have organization knowledge, teaching experience, and preferably direct contract and technical experience in addition to skills, knowledge, and attributes defined for corporate education professionals.

DESIGNING THE WEB BASED CORPORATE INSTITUTE

Once the decision has been made to develop a web based corporate institute and the needs assessment completed and analyzed, executive management should review alternative options in the design of the institute, preferably by someone who has designed, developed, and implemented a corporate institute successfully. Some corporations may choose to hire an external consultant to help them develop the institute. In some cases, if the corporation has an advisory board for the institute, someone from outside the corporation like a customer, a business partner, or consultant, can provide a different design perspective on the institute that adds value to the overall mission.

The institute should be designed based on the corporation's requirements. A written corporate policy on the role of the web based corporate institute is helpful in setting the vision for the institute and the authorization for its existence. Design considerations should address whether the corporation wants a totally web based institute or one with classrooms and labs as well as web based capabilities. In either case, a website should be developed and made available to all employees and managers on the company's intranet. The website would provide information on courses offered, schedules of classes, administrative information, and links to E-learning courses and internal and external resources. Some other design considerations listed here in no special order, should include the following:

- Courses should be organized into a curriculum for separate disciplines…i.e. engineering, finance, marketing, etc., and if it's necessary to provide accreditation, that should be obtained from colleges or accrediting councils.

- An inventory of learning resources should be made available on the institute website. This could include a course catalog.

- Other related missions should be the responsibility of the institute. i.e. college and university relations, tuition reimbursement program, grants to schools if offered by the company, library functions, mentoring programs, and other missions determined by the advisory board. Some of these missions can be accomplished by optimizing the web based features of the web-based institute.

- Depending on the size and profitability of the corporation, an investment must be made to address the requirements. Cost savings and cost avoidance benefits should be assessed.

- The institute should have a director and a small staff depending on the size of the corporation. A network of internal corporate instructors can be developed. Most, if not all, of the products and services of the institute should be made available on the corporation's intranet with the staff providing consulting, teaching services, course development services, employee assistance, and strategic leadership for their respective responsibilities.

- Some corporations may still want the institute to have classrooms and traditional training support capabilities and be located at a physical place in the company. *This is a costly option and a carry over from the days of training departments.* Consideration must be given to enable most employees to access the location in an economical way. The design of the center should include capital equipment needed, material provisions for student employees, breaks, food and beverage services, and accessibility to restroom and classroom facilities that include provisions for persons with disabilities. Classrooms have the advantage of being equipped with the necessary equipment and configurations that support learning. *The option to construct classrooms is not what I would recommend and the corporation needs to decide if it really needs this kind of operation.*

- If the institute will sponsor E-learning courses, connectivity with the information technology organization, the corporation's designated webmaster, and procurement is imperative. Centralizing all training vendor contracts has proven to be an economical and effective way to manage the institute. The institute director should be granted authority to sign all contracts to procure learning solutions and to work closely with the webmaster to design the institute's web site.

- If the institute has international clients or is involved in foreign military sales, consideration should be given to language translation capabilities. International Trade in Arms Regulations, (ITAR) must be understood and followed. Courses in international cultures should become part of the offerings of the corporate institute and maximum use made of web resources.

- There must be an intranet and an Internet website for the institute. If the institute will be marketing its products and services to customers, there must be a separate on line and/or off line financial and accounting system for obtaining payments from customers for services provided.

- If the corporation wants the institute to provide academic credit for courses taken, the best approach would be to partner with a local college or university that's already accredited. Certificates of attendance are sometimes generated and given to employees who attend courses but they have little meaning outside the corporation. Most corporate institutes do not have authority to provide college or university academic credit unless they've contracted with a local college or university which enables them to provide credit for some of their courses. They can seek accreditation for their courses from accrediting organizations and become accredited. The easiest solution is to partner with a local college or university who may sometimes allow for academic credit or for continuing education units (CEU's) to be authorized for some corporate courses.

These are some of the key considerations one should think about in designing the institute. There are many trade offs and options one can consider in designing a corporate institute. Various model options should be presented to top management to enlist their comments and suggestions in optimizing the design of the institute. This process can take several months to a year and involve many presentations to top management. Information on other corporation institutes and initiatives in this area as well as research can help determine the design options that might be useful. Once all of the trade offs and options have been discussed and the decision made to accept one of the designs presented, the next steps can be taken.

DEVELOPING THE INSTITUTE

Stages in developing an institute include developing a centralized budget, staffing, developing the website, and communicating the vision and programs of the web based corporate institute. Once the institute design is approved, the budget must be justified and approved. The first web based corporate institute I developed was at TASC which had multiple locations and geographically disbursed sites throughout the U.S. Only one of its locations had a budget for an institute and it established a location restricted web page which eventually fell into disuse. Information on the location website was not current and often incorrect. Employees at the location eventually ceased to visit the site. Other locations in the company offered classroom training and didn't use the intranet to manage the training and development functions. Some departments at different locations had multiple E-learning contracts with vendors. In short, training and development throughout the corporation was fragmented, decentralized, and uncoordinated. Frequently budget cuts at the locations further diminished training and development programs. Other corporate locations were not easily able to access education program courses offered at any other site. Senior management decided to build an institute at the largest corporate headquarters location and specified it should serve all of the corporation's locations. To justify the budget, agreement had to be reached to eliminate duplicate training facilities, staff, and courses at all locations. This was difficult as many sites preferred autonomy and wanted control over their location's training and development programs. Since The TASC Institute was a corporate initiative and most employees of the corporation were located near the main headquarters of the corporation, a separate training facility was constructed. The location, however, was not in the immediate vicinity of where most of the employee's offices were located and many had to take a shuttle or drive to the location to attend classes. Employees at remote, geographically disbursed locations could not access classroom training conducted at the new facility without incurring travel and subsistence costs. Distance learning began to replace traditional classroom instruction. In the first year of operation, The TASC Institute received outstanding reviews from employees, managers, and from the external partnerships formed with universities and training vendors. Remote locations, too, were enthusiastic in their praise for the institute and it served as a model for the parent company, Litton Industries. Development steps included but weren't limited to:

- *Centralizing* all training and development budgets and consolidating other budgets that pertained to the missions of the institute. The finance and accounting organization is an important partner in this effort.

- *Assuring that the website* was developed and intranet capabilities of the corporation were robust enough to sustain a web based corporate institute. The information systems organization is an important partner in this development effort.

- *Staffing* the institute with capable leadership. The human resources department is instrumental in helping to establish job descriptions and career trajectories as well as to support the institute's mission for human resource development.

- *Developing a Communications Plan.* The communications department is an important partner in helping communicate the vision and programs of the institute to internal and external customers.

The next web based corporate institute I developed was at Veridian. They believed that TASC had a competitive advantage in the asset it called The TASC Institute. Veridian wanted to unify its culture and develop a Veridian Institute. Veridian had no systematic way to train its employees and had grown by acquiring other small corporations which also didn't invest much in training and developing their employees. The CEO of Veridian envisioned a more unified corporation and hoped that a Veridian Institute would help fulfill that vision. Developing anything new is bound to meet with opposition and many managers in the corporation were not enthusiastic about a centralized corporate institute. Some were opposed to losing their overhead training budgets to a central institute and many doubted that a centralized corporate institute could address their unique location requirements. The corporate institute, however, was not developed to perpetuate their existing processes and practices, but as a top down catalyst for developing the competitiveness of the corporation and integrating its various cultures. This top down initiative helped develop the institute quickly but was initially met with resistance and doubt by many in the corporate leadership. But after its first year of operation, the institute was beginning to take hold and started to receive praise from employees and managers. The development process again was to secure a budget this time from an advisory board which also approved the staffing of the institute. The Corporate Chief Information Officer helped to establish an intranet and Internet website for the institute and the Vice President of Communications helped to design and develop brochures, videos, and advertising elements for the institute such as mouse pads, pens, posters, and announcements.

TESTING

After the development process is completed, the web based corporate institute should undergo a test prior to announcing its general availability to all managers and employees. Piloting the on line institute in one division of the corporation is highly recommended and helps resolve any technical issues or problems. Something is always overlooked, and a pilot program is helpful in discovering how the system may be improved prior to fully launching the system corporate wide.

METRICS AND QUALITY IMPROVEMENT

In parallel with the development of a corporate institute a process of continuous quality improvement should be documented and undertaken. This involves measuring the quality of all products and services offered by the institute, benchmarking and comparing institute programs with other companies, determining strengths and acknowledging and improving weaknesses. Maintenance of training records and metrics is important and it goes without saying that *a training management system is imperative*. As mentioned earlier, training management systems are available in the training marketplace that will assist with requirements assessment, skills inventories, training records, gathering of metrics and demographic data, and many can provide a host of other education management tools. To assess the quality of courses provided, student evaluations are generally and routinely completed for most corporate courses. For specific courses that required assurances that the material was absorbed pre and post tests were administered. Post course follow up evaluations with management are often conducted for courses that are job specific. In the case of project management training and business development training some corporations attempted to demonstrate a return on investment to the corporation in the form of improved contract management awards and business wins.

On-going inter-relationships with other staff organizations as well as the technical line and marketing organizations requires that an institute director not only be familiar with the corporation's organization, its products and services, and be an expert in corporate training and development, but also have personal attributes that would enable participation on a wide variety of internal technical, human resource, and business teams. Additionally he or she must represent the company to external universities, suppliers, and customers and have the trust and confidence of senior executive management.

Finally, it is possible with today's technology to develop an institute that is primarily web-based. Management and leadership development courses are available through E-learning. Technical courses and other web-based features of an institute can also be provided on line and thereby minimize the requirements for facilities and staff. However, it is not uncommon to have continued support for traditional instructor led training in a classroom setting. Classroom gatherings are commonly viewed as providing an important social role in disseminating the corporate culture and helping people feel part of the corporation.

LESSONS LEARNED

- A systematic approach to developing a web based corporate institute involves:
  - establishing a baseline and gaining commitment,
  - determining and analyzing requirements,
  - designing and exploring options,
  - developing the institute,
  - testing the system prior to making it fully operational, and
  - adopting a process of gathering metrics and assuring continuous quality improvement.

- The process for developing web based corporate institutes involves establishing a centralized budget, staffing the institute with highly competent professionals, assuring the capabilities of the corporation's intranet, and developing a communications plan. There are different approaches to developing an institute and there are multiple dependencies and multiple constituents that will determine its success.

- Continuous process improvement is necessary for the corporate institute to stay aligned with the changing products, services, and strategic direction of the corporation as well as to make sure it is accomplishing its mission and meeting the corporation's requirements.

- Ideally, a web based corporate institute should not only address training and development needs of the workforce. It should promote and help change the culture of the corporation, be a participant on technical, human resource, and business teams inside the corporation, and represent the corporation to universities, colleges, and training service suppliers.

- A website presence on the corporation intranet and the Internet as well as a strategically focused communications plan is essential for the success of a web based corporate institute.

- Web based E-learning resources can reduce training costs and provide distance learning accessibility to locations remote from the company's primary location.

- Accreditation of courses should be undertaken with a partnership with colleges or universities.

- The corporation should decide whether its tuition reimbursement program is filling its critical skill needs. Tuition reimbursement should probably be a practice rather than a benefit and be a part of the institute's mission. The program should be aligned to fulfill the corporation's most critical professional skill needs.

CHAPTER 6

▼

HOW TO MAKE A CORPORATE INSTITUTE PROFITABLE

Besides focusing on the human resource development needs of the corporation and the organization development needed to integrate newly acquired corporations into the corporate culture, web based corporate institutes can generate a profit for the corporation.

CEO's and executive managers haven't focused on the potential profitability of converting overhead services into profit centers. Most leaders recognize that the education and training of employees is essential in order to improve productivity and stay ahead of the competition and some have established profitable corporate institutes to recover some of their overhead costs. Some corporations are beginning to see the possibility of providing customer education through a web based corporate institute, giving it equal status with other profit oriented business organizations in the corporation, and providing their customers with expanded services.

COST REDUCTION OR PROFIT OPPORTUNITY

Some corporation's unfortunately are quick to reduce or eliminate investments like this when cost pressures jeopardize the business. In many cases, they delegate human resource development cost reduction decisions to human resources and lower levels of management. When this happens, it's a sign that the CEO does not personally champion the cause of human resource development in that corporation. The problem is not that cost reductions are not needed, but the manner in which they're decided and undertaken. Too often I've seen across the board cost cutting enacted so that there's equity in the cost reduction process. This egalitarian way of reducing costs doesn't do much for the high value components of the corporation as determined by the customers, and puts them in the same category as the lesser valued components, from the customer's perspective. I think customers value a highly skilled workforce more than they would value multiple organizational layers of management, or facilities and administration. An in-depth analysis of overhead costs might reveal that facilities costs, inventory carrying costs, capital equipment costs, and building maintenance and repair costs are far greater than the amount invested in human resource development yet if budgets are reduced in all of these functions equitably the customers are probably impacted more by the reductions in human resource development than by reductions in facilities costs. Often labor costs are cited as a reason for cost reductions which often results in layoffs and workforce dislocations. While there are times that draconian measures may be needed to save a business, it is extremely rare when cost reductions and layoffs are smartly directed at reducing too many layers of management in the corporation, executive compensation and bonuses, or focused on reducing costs in areas that don't have a strategic impact on the corporation. ***Instead of cost reduction initiatives, it would serve the corporation to investigate the opportunities to turn the cost center into a profit center.***

In recent years, work has been outsourced to foreign companies resulting in the loss of jobs. This phenomenon is not unique to the U.S. Since people want to buy products and services at lower costs, these pressures have forced executives to seek ways of lowering their overall costs by turning to foreign companies who will provide the same products and/or services at substantially reduced labor costs. But what happens to the dislocated workers who are no longer needed? It's been pointed out that outsourcing often results in increased employment in the U.S. because the company has more financial assets to invest in other areas of the business. But the new work often requires skills and knowledge not possessed by the dislocated workers. *One of the important things a corporation can do is to retrain those workers for employability, either in their own corporation or sell this service to their domestic suppliers, business partners, and/or customers. Retraining and placing workers in other jobs has enormous value for the corporation and profit making potential.* By investing in efforts to increase profitability in areas such as customer education, the retraining and placement of skilled employees into other jobs, and converting traditional overhead services like facilities and human resources into potential profit centers, new opportunities in existing markets as well as new opportunities in new markets can result. Dislocated workers can also be employed by corporate institutes to develop courseware and develop marketing strategies for selling training products and services. Corporations that help retrain and place dislocated employees in other organizations often reap the benefit of building a positive reputation in the marketplace.

FUTURE POSSIBILITIES

In many ways, the future of defense corporations will be determined in part by their efforts to harness technology to solve human problems. *Their future will be determined by how well they apply human ingenuity to use existing and new technologies in new ways to stimulate growth in the marketplace. Their future will be determined by how well they train and educate their managers and employees. And their future will be determined to a great extent by how much more than their domestic and foreign competitors they choose to invest in people and in knowledge in addition to their investment focus on tangible equipment, facilities, and capital.* Now and in the future, training programs must be consistently funded, distributed to those who require training, accessible any time anywhere, modular, interruptible, and structured in multi-sensory formats so that employees can quickly grasp difficult concepts and be able to apply what they've learned. By building a strong training base, the corporation can market corporate institute products and services over the Internet and not only recover some of the operating costs of the institute but also increase profitability. Many corporations have employees in the work force today that lack the skills needed to function in a technologically advanced corporation or are trained to understand the underlying basic science, mathematics, and engineering concepts and their relationship to the technologies of their corporation. This translates into critical technical skill shortages in supplier and customer organizations and an opportunity to provide a needed service. Many would be willing to buy technical skill building services from a web based corporate institute. It goes without saying that a corporation can profit in many ways by establishing a corporate institute in both indirect support to the bottom line and direct support to the bottom line.

INDIRECT SUPPORT TO THE BOTTOM LINE

- The web based corporate institute can be visionary. It can develop strategies by bringing people together for focused problem solving on specific issues. By having a distinguished guest lecture series, new knowledge flows into the company and new, often profitable ideas are stimulated. These ideas can translate into consulting services to the corporation's customers, suppliers, and business partners, as well as ways for the corporation to address contemporary issues.

- The web based corporate institute can prepare the workforce for technology and marketplace changes and can transform the corporation into a technology leader whose consulting services are sought by customers.

- The web based corporate institute can provide a highly skilled technical workforce in times of increasing skill shortages and promote life long learning. For corporations engaged in providing services, this can translate into engineering services contracts with their customers.

- The web based corporate institute can be leveraged in bids and proposals by demonstrating the continuing investment the company makes in keeping its workforce on the leading edge. Well written proposals by skilled and knowledgeable people often translate into contract wins.

- The web based corporate institute can focus on important standards such as the Software Engineering Institute Capability Maturity Model, (SEI-CMM). Support for software engineering, marketing and business development support, and knowledge management initiatives, can lead to profitability. Today, some defense companies cannot bid on software development contracts unless they have a high SEI-CMM level rating.

- The web based corporate institute can help transform the culture of the corporation, help integrate acquired companies into an identifiable, market presence, help people embrace the future, and help support the strategic direction of the corporation. If they are successful they can market this capability to other corporations.

DIRECT CONTRIBUTION TO THE BOTTOM LINE

Today, many web based corporate institutes are contributing to the bottom line by becoming profit centers and selling corporate capabilities, training courses, and services to customers. Customer training can become a mission of a web based corporate institute. Customer training generates revenue by providing training support for the company's products and services. Web-based consulting services and mentoring services can be marketed for profit as well as electronic performance support systems which can range from simple websites to complex search engines and information services. Creating a corporate institute that supports corporate employees and is also a profit center requires keeping separate financial records; one for providing internal courses and services funded by overhead, and one for customer education. Once the financial infrastructure is set up, the company must select what products or services they will sell and determine the development costs for those products and services. Once the development costs have been determined, a competitive price must be offered which often requires a market analysis and determination of which market segment they will serve. Accounts receivable capabilities can be offered on line and by using E-commerce techniques, corporate institute products and services can be procured and paid for on line. Not everyone will be successful in venturing into the training marketplace. Some corporations may believe that all they have to do is to set up an institute, tell their customers that they're selling training services, and that customers will flock to their doors to procure training and consulting services. They may be in for a shock. The training marketplace is highly competitive, has many vendors, is often price competitive, and the development of training products is often labor intensive. *To enter the training marketplace on a for profit basis requires market planning and analysis, an understanding of the market segment the corporation can compete in, and high quality products and services which can be characterized and perceived as unique.*

PROFIT MAKING INSTITUTES

The commercial business segment has been involved in developing profit making corporate institutes for many years while the defense industry business segment has few examples of profit making corporate institutes. Motorola, for example, has a Certification and Learning Services unit of Motorola University which is designated as a for profit group. Motorola University training centers are located in Schaumberg, Illinois and Tempe, Arizona. When I visited their campus in Arizona, they provide training for Motorola employees as well as corporate customers. Their model was to develop a profit-making institute equipped with classrooms and invite customers to visit their location(s). The knowledge, skills, and experience of Motorola employees are provided to customers and business partners to create additional revenue for the company. Walt Disney Corporation established the Disney University to train their employees in the unique skills required by the company. It is one of the largest corporate training facilities in the world and provides a wide range of training courses. Additionally, they established a separate Disney Institute to provide their clients and customers with a variety of training opportunities. The revenue producing institute provides a web page that guides prospective customers through a list of questions. The results are used by the customers to investigate how Disney Institute professional development solutions can be tied to their needs. Another model that combines some of the features of the Motorola and Disney initiatives is Oracle University. This enterprise provides a web page on the Internet that allows customers to view their course catalog, product courses, learning paths, certification programs, online library, self-study CD ROMs, private events, and training locations as well as providing promotional information, customer success stories, and help functions. The web site also enables prospective customers to search for courses in the categories, timeframe, and formats they would like and locate Oracle training locations.

INTERNET WEB PAGE DESIGN

An Internet web page should contain the type of information provided by commercial corporations like Motorola, the Disney Institute, and Oracle University. Disney Institute added the feature of having the customers assess their requirements to determine whether Disney Institute had a solution for them. Oracle University had a list of products, certification programs, and training courses for the customer to choose from. Since access to the Internet web sites is world wide, corporations engaged in work for the government should also make sure that the technical information and training courses that may be sold to international customers comply with the International Traffic in Arms Regulations, (ITAR). Provisions in those regulations define when export licenses are required for transmitting technical information and training abroad. A web based corporate institute can be a profitable asset for a defense industry corporation and its executive leaders who aspire to leading technology and marketplace change on a national and international scale and who aspire to integrating the various corporate cultures that can result from growing through acquisition of other companies.

LESSONS LEARNED

- A web based corporate institute can be configured as a profit center and sell its products and services that will make a contribution to the profitability of the corporation.

- Cost reductions in web based corporate institutes, particularly in training and development are counterproductive to the cause of profitability. Other means of cost reduction should be explored

- Web based corporate institutes represent a disruptive technology that can be leveraged into new market segments and into existing market segments. Visionary executive leadership is required to turn a profit in a highly competitive training marketplace.

- Corporate leaders that aspire to transforming their corporations, integrating different corporate cultures, and leading technology and marketplace change, can use a web based corporate institute to assist them in these efforts. *Successful results in changing the corporate culture can have marketing potential as other organizations may need to address the same kind of transformation.*

- The Internet web design should capitalize on the corporation's strengths and unique core capabilities and to market those capabilities through a web based corporate institute. Defense corporations can leverage and differentiate their unique products and services and apply them to new market segments.

▼

CHAPTER 7

How to Integrate Multicultural Corporations

> While the main focus should be on people, multicultural corporations who wish to integrate and form a new corporate culture need to focus on policy integration, process integration, and infrastructure technology integration.

There's more to integrating corporate cultures than focusing on training and developing people. Executive leadership must set the tone, vision, and standards for integrating the corporation and establish the guiding principles that will lead to the establishment of a single corporate culture. *They must define what "success" means* in the cultural integration process and like systems integrators, focus on systems engineering processes.

TOLERATING DIFFERENT CORPORATE CULTURES

The United States of America was formed by accepting people from all over the world who represented different nationalities, religions, races, and cultures. For nearly two and a half centuries the experiment of accepting and integrating people from various corners of the world and various cultures into a nation have been found to be workable. It's not without problems as we all know, but the guiding principles of making the experiment work included but weren't limited to:

- Developing a Constitution and Bill of Rights that established the country's *vision.*

- Developing a system of constitutional and civil law to establish an interpretation of *justice* in a democratic society.

- Providing the *freedom* for people to learn, understand, and develop *tolerance* for other people's beliefs and cultures.

- Developing a representative government to *serve* the people of the country.

Multicultural corporations would be well served to examine their own "vision", develop their own system of corporate "justice", set a leadership tone of "freedom" and "tolerance", and adopt the principle of "serving" their employees and customers. These principles must be practiced by the top leadership of the corporation so that an example is set that enables employees to believe that their leadership is serious about the vision of the corporation. The leadership must believe in and follow a moral and ethical code to set the example for all employees. As previously noted, IBM established the principles of "respect for each individual", "the pursuit of excellence", and the principle to provide "the best customer service". These principles helped form a strong corporate culture. Any employee who believed that management was not following these principles or had a grievance could use the open door policy allowing them access to upper management to adjudicate their case and resolve their issues. None of the other four defense industry corporations that I worked for had a corporate culture as strong and unified as IBM. Most leaders would agree that a good moral and ethical corporate culture is good for business.

CULTURAL CHANGE AND CULTURAL SHOCK

Nothing lasts forever. As history has recorded, change is the only constant factor that has characterized the evolution of human civilizations and organizations. The human need to band together, form societies, and attempt to bring order from the chaos of life has been studied and documented throughout the ages. Some cultures lasted thousands of years. But eventually, all cultures have a life cycle and ultimately change. Corporations, too, are constantly changing. While some corporate cultures are studied for their uniqueness and contributions to society, other corporations constantly struggle to obtain and maintain a corporate identity. Sometimes the evolution of change is slow and other times it is sudden.

One example of corporate culture shock is IBM. As a multinational corporation, IBM separated IBM operations in Asia Pacific and Europe from IBM in the U.S.A. The parent corporation was headquartered in New York, and international market sales were allocated to IBM foreign nationals located in more than a hundred countries of the world. IBM'ers worldwide identified with an IBM corporate culture and believed that as long as they worked hard they wouldn't be laid off. They believed that they had better compensation and benefits than anyone else and their traditional pensions were secure. When IBM nearly went bankrupt in the early 1990's, many of these cultural elements and beliefs underwent drastic change which resulted in a cultural shock to employees worldwide. Dramatic and sudden economic and social changes caused an earthquake-like shock to the IBM population. Cultural shock occurs when a corporation gets in trouble and abandons its basic beliefs and practices. It occurs when one corporation acquires another corporation or divests itself of a business unit segment. There is immediate uncertainty among employees of the acquired corporation as to how the change will affect them. I recall how I felt when Loral acquired the IBM Federal Systems Company. First there was disbelief that IBM was actually selling a profitable segment of the business to another corporation. Then when the reality sank in came depression with the loss of my identity as an "IBMer". Anxiety and anger followed when there was little initial communication as to how the sale of the business unit would affect me and other employees and the rumor mill churned out misinformation. Anger was directed at corporate management who had failed the corporation and in the pursuit of abandoning the IBM cultural beliefs were looking after their own interests. Corporate executive management was responsible for the catastrophic events but they pointed the finger at everyone and everything except themselves. Finally, after there was a stream of information to employees about the divestiture and about the impact to them, and Loral management began a series of town meetings to welcome the former IBM employees into the new corporation, I gave up my IBM badge, acquired a Loral badge, and started to adapt to the new corporate culture. The healing process had begun, but the hard lessons learned from the experience changed my view of corporate America forever. Corporate culture shock is revolutionary, sudden, and leaves lasting impressions. Corporate culture change is evolutionary, slow, and doesn't necessarily leave lasting impressions.

METHODS CORPORATE INSTITUTES USE TO HELP INTEGRATE MULTICULTURAL CORPORATIONS

Some corporations use web based corporate institutes in the following ways to help integrate acquired companies into their corporation:

- Corporate institutes survey the training and development needs of the acquired corporation and blend them into the solution for the annual internal training and development plan.

- Corporate institutes sponsor executive and management on line and in-person integration meetings to resolve strategic problems and issues.

- Corporate institutes educate management on the latest organization development practices and how they can be applied to integrating the corporation.

- Corporate institutes provide on going communications on their web site to inform employees of best practices and the progress of integrating new corporate tools, methodologies and practices to improve customer service.

- Corporate institutes help executive management explain, communicate, and when necessary redefine the corporate culture.

- Corporate institutes provide "one stop shopping" for knowledge and skill acquisition, knowledge transfer, and access to performance improvement information.

TIME AND PERSPECTIVES

To develop a perspective of how to integrate multicultural corporations, it is helpful to have a look back, a look into the present, and a look to the future. By looking back we can benefit from lessons learned. By looking at the present we can see trends that can assist in decision-making, and by looking into the future, we can develop plans based on current trends to help chart a course that may be the best for a corporation.

Looking Back

Experience gained from historical and sometimes dramatic corporate change can give one a more objective view of what happened and put a different perspective on the hard earned experience. While lessons learned are usually found at the end of the chapter, here are ten lessons I've learned from looking back on some past experiences:

- First lesson learned is that not all corporations, even those in the same industry, are alike. Each develops a corporate culture that differs from other competitors in the same industry. Each reacts to change differently.

- Second lesson learned is that not all corporations want to integrate different corporate cultures they've acquired into one corporate culture. Many corporations don't recognize that they have a corporate culture and couldn't describe it if they were asked. Many corporate leaders don't see the value of having a single corporate culture.

- Third lesson learned is that those who have successfully established a strong corporate cultural identity often are pleased with the results and accomplishments. Many have excellent reputations with their customers, high morale among their employees and are readily recognizable in the international marketplace. They don't want to change and strong external forces often force them to change

- Fourth lesson learned is that corporate acquisitions and mergers can take a toll on employees, managers, and customers. The former corporate culture is no longer applicable in the new corporation. Employee morale suffers, managers are often confused by different procedures and policies, and customers begin to wonder how the change will affect them.

- Fifth lesson learned is that corporations who successfully integrate other corporate cultures into their organization follow a process and use techniques that include:

  - frequent communications with employees on what the change means to them,

  - frequent communications and press releases to assure customers that the new corporate acquisition will be good for them,

  - the use of corporate web based institutes to provide needed employee services

- organization development techniques,

- the use of corporate symbols. Lockheed Martin, for example provided star pins for its employees and Veridian provided a gold pin with the Veridian corporate name on it, and

- A dedicated strategic plan for integrating the corporation and comprehensive support to the newly transferred employees to reassure them of their benefits, compensation, and their value and importance to the new corporation.

- Sixth lesson learned is that a corporation's culture is a valuable asset. It helps to have a branded logo that customers and employees can identify with and provides reassurance to employees and managers that the corporation is not there for the short term.

- Seventh lesson learned is that you need talented people to lead a web based corporate institute and to support the integration of multicultural corporations. This is not a job for human resource generalists or ad hoc committees.

- Eighth lesson learned is that a corporation's intranet can be optimized to streamline policy, practice, and procedural changes as well as to provide. distance learning and E-learning technology to play an important role in corporate training and development, particularly as they relate to information technology training and the integration of infrastructure technologies.

- Ninth lesson learned is that change is ever constant and change leadership is better than managing the impact of sudden change. The past is not necessarily prologue to the future. Disruptive technologies, dynamic changes in the international marketplace, and demographic changes in the population all have multi-year, uncertain consequences for corporations. Corporate executive management must stay alert and make wise long term decisions to avoid cataclysmic events.

- Tenth lesson learned is that the Internet and the World Wide Web are being improved every year and new technologies are enabling the established corporate universities and institutes to evolve into efficient and cost effective web based institutes. To survive, the corporate institute must adapt to organizational needs, technology change, and marketplace changes. It must maintain a perspective on how new technologies are reshaping the institute's design and mission.

To avoid the errors of the past, corporate leaders must examine the lessons learned from past mistakes so as not to repeat them in the future. Past mistakes might include but not be limited to,

- Lack of attention to financial realities like excessive overhead costs growing out of proportion to revenue.

- Insufficient or lack of strategic planning and insight into changing market conditions and technology.

- Lack of understanding of corporate cultures and how they impact the corporate markets, employees, and customers.

- Lack of a robust intranet capability and application of new technology to improving the corporation's technical infrastructure and lower its overhead costs.

- Lack of tolerance for other corporate cultures and lack of appreciation as to how an acquired corporate culture might benefit the parent corporation.

- Insufficient attention to customer concerns and impacts of corporate culture changes.

Looking at the Present

Many corporations today are struggling with how to better manage their overhead costs following a major acquisition. Problems and issues regarding integration of multicultural corporations are often erroneously delegated to human resources. Technology is often overlooked to help integrate the corporation. Some of the common place technologies in use today can help integrate a corporate culture. For example today's cell phones can connect to the Internet, have cameras, and can transmit pictures, text, as well as audio messages. Personal data assistant devices include more and more functional capabilities. Small video cameras can connect to virtually any computer. Internet search engines can provide access to encyclopedia-like web sites to find answers to many questions. This vast improvement in communications can help employees adapt quickly to their new corporate identity. The explosion of software applications, electronic communications devices, and new ways to access world-wide information is impacting the way we all live and think. Other wireless devices also increase the accessibility of information and knowledge. *The challenge is to find ways that these new technologies can be employed to achieve multicultural integration.* Some observations from viewing the present technology, marketplace, and demographic trends would be that:

- There's an explosion of new communications technologies worldwide and new ways are being found to optimize E-commerce and web technologies.

- The marketplace is global and creates new business opportunities, new competitors, new products, new services and corresponding need for people who understand how to capitalize on these changes.

- Corporations have critical skill needs in technical disciplines that are not being met by new college graduates. In addition to technical skill needs, business leadership, communications, and financial knowledge as well as positive personal attributes are required.

- Corporations must train newly hired and long term professionals if they are to retain them.

Looking at the Future

When a corporation is impacted by major waves of change the corporate vision of future markets and customers is fractured. Today there are many waves of change that cloud the landscape making it almost impossible to determine with any degree of certainty what trends may lie ahead socially, politically, economically, scientifically, and in any other field. IBM was the most admired corporation in the mid 1980's. Xerox, Eastman Kodak, AT&T and other highly respected corporations were also highly successful during this period. Then came the 1990's, the Internet, and the World Wide Web. New corporations that no one had ever heard of before like Microsoft and Intel emerged seemingly from nowhere and became highly successful, multinational corporations like the older corporations in some ways but entirely different in other ways. The new corporations took advantage of changing technology. Xerox, Eastman Kodak, and AT&T are still trying to transform themselves as a result of major waves of technology change in their marketplace. The speed of change is breathtaking. One can expect this kind of sudden change in the future and the impact on the corporation will be unpredictable unless executive management sees the trends and adapts to them. When IBM sold its Federal Systems Company to Loral another wave of change was descending on the defense industry which resulted in numerous divestitures, acquisitions, and mergers. Corporate culture shock rippled through defense industry corporations trying to adapt to the sudden changes. Loral was a small corporation by IBM standards and was more widely known in the government defense arena than in the commercial international marketplace. Two years after the purchase of IBM's Federal Systems Company, Loral sold most of it to Lockheed Martin and top Loral executives were offered offices at Lockheed Martin's Corporate Headquarters in Bethesda, Maryland. Most of them never accepted the offer because soon after the sale was announced, Lockheed Martin spun off L3 Corporation. Lockheed Martin was formed in a merger between Lockheed Corporation and Martin Marietta. The wave of mergers and acquisitions continued as Lockheed Martin acquired business units of General Electric, General Dynamics, and Loral's defense systems business units. TASC was acquired by Litton Industries and Litton Industries was acquired by Northrop Grumman all taking place in a few years. Veridian acquired Signal Corporation and then a year later was acquired itself by General Dynamics. ***Defense industry and non-defense industry corporations are finding that they must transform themselves to stay competitive. To do this they must learn to apply new strategies and concepts faster than their competition.***

LESSONS LEARNED

- Mergers, divestitures, and acquisitions are not confined to the defense business. Banks, automobile companies, and virtually all industries have seen an acceleration of major mergers and acquisitions in the past decade mostly brought about by marketplace changes, globalization, and the rapid pace of technology change. From the standpoint of employees, most corporations have unfinished business in completing these acquisitions.

- Throughout the 1990's and into the 21st century, the development and implementation of new technologies proceeded at high velocity. The United States demographics have changed. The baby boom population born after World War II is aging and is nearing retirement causing stresses on pension plans and health benefits. The U.S. became a target of terrorism and went to war to defend itself, and wave after wave of historic breaking news has stunned the average American. The world itself is now interdependent and interconnected. It is difficult to predict what may happen tomorrow, but we'd better learn to spot trends and strategically plan for what we anticipate will occur. The indicators are there, but the management mind set has not caught up to the realities of change.

- We can't predict what will happen tomorrow but we may be able to predict with some confidence that the rate of technology change, globalization, and marketplace change will probably continue and there will be a never ending need for highly skilled, educated people.

- The ten lessons learned in this chapter under the topic "Looking Back" may help integrate a multicultural corporation.

- We live in times of great change and great opportunity. The new generation of Americans must learn to look ahead and make wise decisions concerning their own future. The defense industry cannot be maintained; it must be renewed and reinvented.

CHAPTER 8

▼

WHAT ARE SOME OF THE DIFFICULTIES IN DEVELOPING AN ON LINE INSTITUTE?

The only difference between where you are in the corporation and the top business leaders of the corporation is that they got there first. While some corporate managers and leaders may have extraordinary insights as well as good fortune, they must learn to lead, encourage, and support good ideas from everyone in the corporation. The biggest obstacle to developing a web based corporate institute is bureaucratic and stodgy top management.

Developing a web based corporate institute takes will, vision, time, patience, and perseverance. The process to design and develop an institute is not difficult to follow. Most of the difficulties lie in getting a focus on the following areas:

- Consistent funding.

- Quality of the services offered by the institute.

- Individual learning styles.

- Lack of a robust intranet.

- Lack of experienced leaders for the institute.

- Lack of commitment.

- Delegation of authorization and accreditation decisions.

Let's examine each of these elements in turn.

CONSISTENT FUNDING

In most companies, annual budgets are developed and submitted for review and approval to senior management by each organization. The previous year budgets are reviewed and compared with the new budget submittals and usually there's an increase in the amount requested due to increasing supplier costs, salary increases, new programs, and other issues. Rarely, does the enterprise compare its budgets with other company's. The American Society for Training and Development has a benchmarking program where companies voluntarily submit their training budgets and complete surveys as to how the budgets are spent. Training Magazine, too, annually publishes a survey of training and development programs comparing company expenditures to the previous year. In my experience, smaller companies tend to spend in the neighborhood of 1% of their annual payroll, (minus benefits costs), the average company spends around 2% of payroll, and larger companies spend upwards of 3% of payroll on their training and development programs. The expenditures, however, vary from company to company and industry to industry every year. Sometimes there's an emphasis on leadership development and sometimes an emphasis on information technology and other areas. The budgets often include the costs of the training staff, development, design, and delivery of training programs and usually do not include the cost of student labor, (employees attending classes during work hours). It is the student labor costs that are often the major concern of executive management in the defense industry. When employees attend classes on company time, productivity is lost but the employee's labor costs are still incurred. Consequently, some corporations have chosen to provide training part on and part off company time or totally off company time. Employees understand this approach but often are critical of the corporation and feel that the corporation doesn't want to invest in them and that they're left alone to upgrade their skills and knowledge on their own time. E-learning courses on the Internet or intranet are accessible any time and at any place. Consequently some of the burden of student labor is avoided if the employee decides to access a course during lunch time, while on business travel, or just before or just after work hours. Costs of E-learning contracts are often justified based on the high volume and diversity of courses they offer, the need to stay current in information technology and other areas, and the cost avoidance of classroom courses and student labor. It's important to have consistent and well planned funding for this initiative so that the web based corporate institute will be able to meet its requirements.

QUALITY OF SERVICES

Most training and development organizations obtain student evaluations from each of the training courses they provide but don't normally evaluate the other services provided by the organization. Training course evaluations tend to focus on the quality of materials, effectiveness of the instructor, and logistics of the course. In general, the quality of on line courses is not reviewed as frequently as classroom courses but E-learning vendors review the quality of on line courses and improve the quality of their products. Courseware developers in the business of selling on line E-learning realize that the quality of the learning experience will determine future sales of their products. The use of audio, graphics, video, and simulations has improved the quality of courseware and addressed differences in learning styles. Many employees want continuous access to high quality learning opportunities in order to maintain their knowledge and skill levels and acquire new knowledge and new skills to remain competitive in the marketplace. Courseware provided on line today, while getting better, has varying degrees of effectiveness and varying degrees of quality. The best programs are offered by training vendors. E-learning contracts are often based on the projected number of employees who will access the on line training. Increasing the number of employees who access E-learning courses will lower the cost per employee. Consequently, vendors often provide marketing assistance to corporations who purchase their products and help the corporation sell E-learning to their employees. Contracts with E-learning providers can be expensive and the company needs to assess the cost of quality versus the cost of ineffective training provided at lower costs.

INDIVIDUAL LEARNING STYLES

We have all undergone the experience of public education. The teacher is in front of the class and students passively absorb and try to retain what is taught. Once in awhile there's interaction between the students and teachers. This model of instruction was developed from the days of the industrial revolution and paralleled what was expected in the workforce at that time. The teacher played the role of the boss and was the unquestioned authority in the classroom. Teachers were there to impose order on the learning process, teach the subject matter, be listened to and obeyed. Parents generally supported the authority of the teacher. School started early in the morning (the same as the job); when the bell rang you had recess (coffee break on the job); when the bell rang again you had lunch (lunchtime on the job). At the end of the day you caught a bus ride on school provided transportation and went home. *Today, classroom instruction cannot keep pace with the higher volumes of topics to be learned and discussed. The industrial model of instruction no longer fits. Yet corporations still use this traditional model and resist change.* The work environment has drastically changed. Advances in education psychology and learning theory have shown that each of us develops our own method of learning. Some learners absorb and learn new material better if they have hands on practice on what they need to learn. Some learners absorb and learn new material better if they like the subject and the instructor. Some learners prefer to read, study, discuss, and recognize patterns in the new material that they need to learn. Not all learners are motivated by the same factors and group instruction often is not effective unless different classroom techniques are employed such as role-playing, team projects, labs, workshops, games, electronic self-study tools, Internet access, and multimedia presentations of material. Training professionals know what it takes to have a successful learning experience and the preferred method today is to provide a blend of learning options. With the rapid advances in technology, the industrial classroom is being replaced by a web-based electronic learning medium that adapts to individual learning styles. It is important to remain focused on this item because learning and applying what is learned is the main reason for investing in a web based corporate institute's training and development program.

LACK OF A ROBUST INTRANET

Most discussions of intranet capabilities revolve around bandwidth which is the transmission capacity of the connection to the Internet or intranet. More bandwidth is better because bandwidth enables the use of multimedia, audio, and video over the network. On line education offers collaboration between people and institutions but this cannot occur without the support of network administrators and a robust corporate intranet and Internet. There may be some techni-cal, political, and social hurdles to overcome. For example, today's courseware offers video, animation, graphics, audio, and text which require higher band-widths. Upgrading the robustness of a company's intranet can be costly and the justification required to upgrade the intranet has to be based on more than offer-ing high quality courses. Other company departments can benefit from a well planned intranet. For example, improvements in the intranet capabilities were obtained in the corporations I've worked in by teaming with marketing, finance, human resources, and procurement all of whom wanted to use the company intranet for their own purposes to enhance their products and services. A robust intranet allows for quick responses, chat rooms, multimedia, and other forms of collaboration. Assuring network reliability is a first step in being able to provide a web based corporate institute. While not always necessary, video streaming and audio features are desirable. Some learners prefer audio based training. Some pre-fer video based training. Generally, most learners prefer text, audio, and video capabilities in on line courseware. Employees and customers often judge the effectiveness of the corporation's learning programs through their experience on the web. If the system experiences downtime, video or audio difficulties, or there are disruptions and interruptions to the on line learning experience, the quality and effectiveness of the program suffers. Other potential technical hurdles may be the configuration and capabilities of the individual workstations. Some may not have the appropriate software applications or workstation configuration to sup-port accessing on line courses. The corporate institute web site should provide instructions as to what technical configuration is needed to participate in on line E-learning courses. Some learners may require assistance in order to accessing on line courses from either the intranet or the Internet. Frequently asked questions about technical matters and a help line should be available on the corporate insti-tute web site.

There are often political and social hurdles to overcome. The political and social hurdles often involve non-technical issues such as pre set attitudes, intellectual property, and compensation of instructors. Some people have made up their minds based on past experience that E-learning has too many issues to be useable. It has been my experience that people will oppose corporate institutes, distance learning, and web based E-learning because they don't believe in it. In some cases, they oppose the concept without ever having tried to take a web based E-learning course. Intellectual property is sometimes an issue with instructors believing they "own" the course rather than the corporation owning the course. If an instructor is on line reaching more people than can be reached in a classroom some believe they should be compensated more than a traditional instructor. Overcoming strong political and social pushback to establishing something new has always been a challenge. Courseware is improving but still has a way to go to address the special needs of people with disabilities. People with audio or visual disabilities must be accommodated in corporate training and development programs.

Most corporations strive to provide the best capital equipment affordable to their employees and they are improving their intranet capabilities every year. Equipping everyone with the latest personal computers and communications technology is seen as an important step to improving productivity and speed in the marketplace. These are all important focus areas because management must often intervene and take action to solve problems as they occur to avoid frustration and dissatisfaction. Management should:

- Listen to all employees and understand their concerns and complaints.

- Try to find solutions for all problems encountered.

- Seek out and rely on business partners to assist in providing solutions to issues.

- Support the acquisition of courseware that assists people with learning or physical disabilities.

- Work closely with the webmaster and the information systems organization to improve the intranet and solve immediate problems encountered.

- Communicate with their internal and external customers. Let them know what the corporation is doing to improve the quality of its products and services.

LACK OF EXPERIENCED LEADERS FOR THE INSTITUTE

The question of who should lead the effort to develop the corporate institute and provide a web-based institute capability can be difficult. Should the institute director be?

- An educator or communications professional with an advanced degree in adult education familiar with academia, learning theories, teaching, and education administration.

- A project manager with an advanced degree in business familiar with the business organization, finance, the company's products and services.

- A chief information officer with an advanced degree in information technology who's familiar with the company's intranet, technologies, and processes.

- A marketing executive with an advanced degree who's familiar with selling the merits of the institute, and the importance of communications, and advertising.

- A generalist executive with an advanced degree who's familiar with all of the above knowledge, skills, and attributes.

Notice the reference to advanced degree in all of the above criteria. In order to be credible, one must have credible credentials and be able to establish quick rapport with internal and external colleagues. More and more corporations are turning to executive search agencies to attempt to find a suitable candidate who has all of the above experience and a multi-disciplined background to be able to lead a corporate institute. Rotational assignments as the institute director can serve as a development assignment for senior leadership positions since the scope of relationships that an institute director has crosses all organizational functions and involves high level contact with external business and academic leaders. It is essential for the institute director to have outstanding communications and interpersonal relationship knowledge and skills.

LACK OF COMMITMENT

Perseverance and commitment are leadership qualities that, if lacking, will doom any effort to develop and maintain a corporate institute. This is not a one time "fire and forget" kind of effort and requires an on going time commitment and "buy in" from the executive leaders of the corporation. The corporate CEO should spend up to 30% of his/her time reviewing succession plans and developing future leaders, communicating the status and plans of the corporation to employees and management, and chairing the advisory board of the web based corporate institute. The remaining time should be spent with customers, the board of directors, external business partners, and strategically managing the corporation. The institute director should spend at least 30% of his/her time with customers and the rest of the time on the day to day management of a network of suppliers, supporters, and staff. If the leadership is not committed to making the enterprise successful, it is doubtful that others will be inspired to make it successful.

AUTHORIZATION AND ACCREDITATION

The authorization for the web based corporate institute must be documented in a policy statement for the corporation that clearly defines the authority delegated to the institute director and missions and purposes of the web based corporate institute. Too often, this is overlooked and the authority of the institute director vis a vis other managers and functions in the corporation are left unclear.

Most adult learners today would like to receive academic credit for their time spent in corporate classrooms. This poses the opportunity for the corporate institute to either seek accreditation for its own courses some of which may be very specific to the company and therefore difficult to obtain accreditation or to partner with a local college or university that already has accreditation. This sometimes involves modifying the courses to add more academic theory to the content, thereby increasing the length of the course(s). Increasing course length means more time the employee has to spend away from the job and doesn't always add to the learning experience. Colleges and universities view a partnership with a corporation as an opportunity to enroll corporate students and thereby gain additional tuition and alumni when the students graduate. This also is an opportunity for them to seek research grants, partnering opportunities, and to solicit funds from the corporation. As mentioned earlier, the path of least resistance is probably to partner with a college or university to gain accreditation in those specific areas that will benefit the corporation.

LESSONS LEARNED

- Seven factors were summarized as difficulties in establishing an on line corporate institute including but not limited to consistent funding, quality of services, individual learning styles, lack of a robust intranet, lack of experienced leadership for the institute, lack of commitment, and authorization and accreditation.

- The excitement and joy of being responsible for a web based corporate institute is knowing that there are many more social, political, and managerial difficulties to overcome and managing them as the issues arise.

- An institute director must have direct business experience and the knowledge, skills, and attributes to lead the institute. Rotational assignments can broaden the experience of future leaders. Outstanding communications skills and interpersonal skills are essential.

- Obstacles exist to be overcome and perseverance and commitment are essential to the success of the institute.

▼

TRENDS THAT SUPPORT A WEB BASED CORPORATE INSTITUTE

> The convergence of several Internet based technologies allows for increased interpersonal collaboration, knowledge acquisition, and real time learning without regard to distance or time.

Unprecedented electronic connectivity among people of divergent cultures allows employees to collaborate on business proposals, allows company's to outsource labor intensive work to foreign and domestic corporations and lower their costs. The technology is here, today, to allow for the development of a web based corporate institute.

INTERNET TRENDS

The trends in using the Internet seem to be finding ways to optimize Internet technologies and use the World Wide Web to collaborate, facilitate commerce, and share knowledge. In the last decade, more and more uses have been found for the Internet and technology has been applied to improve Internet provided services. For example, new applications such as anti-virus security programs and software programs allow people to purchase services or products over the Internet and pay for them on line. Internet Service Provider, (ISP), support services have been improved and expanded. Encyclopedic websites provide the opportunity to search for and share information on a world wide scale. E-commerce is now common and many people shop for books, clothing, and virtually any other product over the Internet. E-Bay allows people to use the Internet to buy and sell a variety of products. There's a trend to find ways to optimize technology to deliver education. The Internet is now widely used by colleges and universities to provide undergraduate and graduate degree programs and other courses. Expanded use of these on line learning services indicates that the subscriber customers seem to like it.

Defense corporations need to find ways for their employees to access a world wide ocean of knowledge. Today, the Internet provides a vehicle to search world wide for information on virtually any topic. It is a universal library, shopping mart and highway for information exchange such as the world has never before seen. It allows for people in the U.S. who need help in fixing their home computer problems to talk on line and chat with technical people in India or other countries who provide that service and help them fix their computer problems. It allows for corporations to outsource services to international companies and collaborate with them to deliver products and services to any market in the world. Daily, people use the Internet to communicate with colleagues, family, and friends, get the weather report and news, send post cards, seek travel information, book hotel and travel reservations, obtain health advice, and find local, national, and international resources from their home computer. And that isn't all they can do with the Internet. Planned upgrades to the Internet will enable every cell phone to have an Internet address, and eventually provide Web-enabled sensors that will someday be embedded in our homes, automobiles, and communities. The Internet has transformed the way people around the world live. It has opened communications channels and provided alternative ways to obtain information people need. It is not out of the question, therefore, to see the possibility of web based corporate institutes being founded in many countries of the world. The World Wide Web based institutes could evolve over time allowing people to take courses from anywhere in the world. Web based training and development is still in its infancy and it's hard to believe that at the beginning of the 1990's, few people knew of its capabilities and existence.

OTHER INTERNET APPLICATIONS

Public school classrooms today are less automated than your home. Public education has seen slow change over the years and hundreds of millions of dollars are spent annually in this segment of our society. I think most of the poor results we see are caused by a lack of visionary and committed leaders who don't want to take the risk of trying something new. We're seeing some progress in higher education but not enough progress in public education. Closer collaboration between public schools and higher education is one way to improve public education. The development of science and technology secondary schools has also resulted attracting gifted students to technical careers and improving the quality of science and technology education. But the biggest impact to public education improvement is yet to come and will undoubtedly be the Internet. Imagine that someday, children will be able to learn from any teacher, anywhere in the world and teachers will be able to access continuous learning in their subject expertise on line, anytime, anyplace. Other Internet applications will soon affect associations, public health organizations, hospitals and other public institutions. They should start now by forming and developing their own web based institute and sharing courses and instructors on line. Distance learning and telemedicine in the medical profession is steadily gaining more acceptance. Technology has always been used when it has been made available, when it is reliable, and when it helps solve problems.

OTHER TRENDS

Many corporations today are transforming their training departments. While technology and the need to reduce costs are driving forces that are leading this change, other factors are also visible. Many corporate leaders are unsure of the value of traditional training departments. Many corporate leaders are looking for better, different, and more cost effective solutions to providing on going training and development. Some corporations have formed corporate universities to serve to attract and retain valued employees. And some corporations have begun to see the profit making possibilities in the training field. All of these forces are transforming traditional ways of organizing and delivering human resource development services.

LESSONS LEARNED

- The convergence of multiple technologies has enabled a more robust Internet and web-based menu of options. Web based computer training, performance support tools, asynchronous and synchronous delivery systems for on line training which are widely used now will be improved upon in the near future.

- An expansion of on line distance learning full degree programs and courses have occurred in the past five years, particularly in higher education.

- While the technology exists to create an on line corporate institute, most corporations have been slow to explore improvements in this area. Many are becoming interested in transforming their training departments and are experimenting with different forms of organization.

- The belief, today, is that there will always be a need for traditional classroom instruction and that technology assisted training replaces instructors. Since knowledge resides in people, it is clear to me that instructors will not be replaced by web based learning. But, their classroom will not be a traditional one. It will be on the Internet or on corporate intranets.

- We are all being challenged by the acceleration of technologies that better enable access and sharing of knowledge and information. Today's Internet is also starting to challenge the assumption that we'll always need traditional libraries and always need the industrial age classroom.

CHAPTER 10

▼

CONCLUSION

> *It is my belief that the corporate training departments and corporate universities of today will be replaced by the web based corporate institutes of tomorrow. Furthermore, corporate institutes can dramatically improve corporate performance and help senior leaders complete "unfinished mergers" and help transform corporate cultures.*

In medieval times, stone castles dotted the countryside of Europe. They attracted itinerant merchants and craftsmen who would wander from castle to castle selling their labor, products, and services. When wars occurred between the castle rulers which resulted in one castle taking over another, it was a common practice of the time to plunder the castles treasures, take prisoners, and scatter the occupants of the enemy castle throughout the countryside. Today, corporate castles of steel and glass dot the countryside of many modern nations. Itinerant professional college students and workers wander from corporate castle to corporate castle seeking employment much as their ancestors might have done during the middle ages. When modern economic wars between castles occur and one corporate castle takes over another corporate castle, the assets of the acquired corporation are taken over and the occupants of the former corporation are either incorporated into the new corporation or laid off. Human behavior hasn't changed much over the centuries. As in times past, the castle rulers profit by the takeover. The only people that don't profit are the employees. The merchants of medieval times have become the corporate executives of modern times. Their quest for increased wealth and power has resulted in an imbalance in the concentration of wealth and the assurance that further wars and takeovers will result in the future. Perhaps it's the nature of capitalism that prompts human behavior to focus only on the accrual of personal wealth rather than also focusing on the accrual of wisdom. Education has attempted to capture the successes and failures of humanity throughout history so that future generations won't make the same mistakes as their ancestors. But every time a new baby is born, it enters the world just as primitive as the first born human child and must learn more and more to survive. Some of the lessons of history are often lost in our journey into an uncertain future.

The development of web based corporate institutes supports my belief that corporation's must pay more attention to the human side of technical progress and invest more heavily in educating and training the workforce and future corporate leaders. I believe that today's corporate training and development organizations will transition into more centralized, web based entities that:

- Will enable instructors to broadcast their knowledge and experience in web casts to their classes which will be dispersed nationally and perhaps globally. Other organization staff functions like libraries, databases of organization knowledge, and initiatives to blend newly acquired corporate business units into a new corporation can also be incorporated in the mission of the web based corporate institute.

- Web technology advances have now made it possible to develop a web based institute on the company's intranet with a link to the Internet. Cell phones with Internet and World Wide Web access offer additional options for accessing and sharing knowledge.

- Different information technology infrastructures can be blended into one corporate infrastructure with the support provided not only by corporate information technology departments, but also by web based corporate institutes.

- The process for developing a web based corporate institute takes into account that there are many design possibilities available to corporate management to enable the corporate institute to provide better access to information, training, and customer education at lower costs. Other staff functions can also use the corporation's intranet to optimize their products and services.

- Based on my experience in developing major corporate training centers and corporate institutes that optimized the use of the corporations' intranets, the time has come to look into the future, capitalize on new technologies and recommit corporate leadership to human resource development.

A web-based corporate institute can act as a catalyst to better integrate people into the corporation and help them acquire needed skills, knowledge, and attributes. It can also serve to humanize the corporation and prepare corporate leadership for the challenges of the future. Additionally, operations that can easily be migrated to the institute's intranet web-site include, but aren't limited to:

- Administrative information, (benefits, schedules, announcements, policies, organization points of contact, etc.)

- Marketing information, (brochures, services and products for sale to customers, billing procedures, sales points of contact, etc.)

- Corporate acquisition and merger information, (intranet links to benefits, corporate organizations, etc.)

- E-learning courses, (access to technical and non-technical courses, enrollment procedures, course evaluations, tests, etc.)

- Access to databases of corporate knowledge and performance support resources, (bid/proposal information, systems and software engineering standards, project management and technical lessons learned, quality assurance, etc.)

As the web based corporate institute grows and matures, additional missions can be undertaken and performance support aides can be migrated to the intranet through the corporate institute web page.

OVERCOMING RESISTANCE TO CHANGE

Here are some tips that may help overcome resistance to change based on some, but not all of the objections often raised about a web based corporate institute:

- ***Not all training and development courses can be delivered on line.*** My belief is that most, if not all, courses can be delivered on line with "wrap around" human interventions. For example, a course in developing team work is normally delivered in a classroom. Today, on line simulations and avatars can act as a team and be managed by a leader. The leader experiences all of the normal problems of gaining teamwork by managing the avatars in the simulation. Poor performance, absenteeism, low morale, etc. are all simulated in the development of a product or service and the leader required to make decisions. Each decision made is didactic and influences the outcome of the on line exercises. An instructor provides a "wrap around" for the learner, helping them to get started in the on line simulation and summarizing the exercise at the end of the simulation. Pre and post tests can also be administered on line as well as the collection of course evaluations. Some corporations specialize in delivering on line courses in human relationship and leadership development areas and provide the courses on line.

- *Not all individuals can learn on line.* My belief is that on line learning can be more effective than traditional classroom learning and there are many studies that have supported this belief. New employees entering the workforce have far more computer expertise and on line experience than senior employees and have become in many respects, "virtual learners". We tend to be more comfortable with what we've experienced in the past as learners and most of us have experienced the industrial model of a teacher in front of the room providing instruction. But today, younger employees have become so familiar with interactive multimedia, computer simulations, and video learning that it is difficult to retain their attention with traditional classroom lectures. Since we all learn differently and at our own pace, lockstep instruction won't be of much help to us in the future. Small learning capsules, delivered on line, interruptible, modular, bookmarked, and in multimedia format, available when and where we need the knowledge, and web accessible, is most probably the wave of the future. With some work by the student and imagination by instructors, computerized simulations and games can be converted to educational experiences to optimize on line learning experiences for everyone.

- *Not all material to be learned is available on line.* My belief is that there is some truth to this argument. Quality courseware is available today for some subjects but not for all subjects. Individuals in the corporation who possess this kind of knowledge often don't have the time to become instructors and share it in a classroom setting either. Providing the material on line in web cast seminars to the people who need it is probably a better course of action. Not all material has to be provided on the corporate intranet. Course material must be available, of high quality, and modularized to be effective. Instructional systems design methods, once a bureaucratic maize of time-consuming tasks, can be modified to structure materials so that employees can learn from them. Information is not knowledge anymore than telling is teaching. A talented corporate institute staff can resolve most of these issues and help in accessing a world wide pool of available and new material.

- *The corporation intranet and employee workstations don't have the capacity to support an on line corporate institute.* There is no doubt that the corporation has to have a robust intranet to support employees. Many employee workstations may be configured differently and may not have the capacity to handle some courseware or web casts. This may be a show-stopper for establishing an on line web based institute. But, my experience with large, medium, and small corporations has been that the intranets have the capacity to support an on line institute and workstation computers are being modernized. Once the requirements are understood, other functions in the corporation may help justify the intranet capacity and technical capabilities needed.

- *It's easier to schedule a group of employees for a class rather than to get them all together in an online web cast.* This may be true in some corporations. Scheduling is never an easy task and even when a class is scheduled, many still aren't able to attend. There's also the cost of time away from the individual's work location, travel to the classroom, etc. to consider. My belief is that fewer, if any, classrooms are needed in tomorrows workplace. Chat rooms can offer the social action needed by some to learn, and it's just as easy to schedule a web cast as it is to schedule a classroom event.

- *When one corporation acquires another there is no need to retain the former corporation's overhead staff or training capabilities.* It is my belief that there is probably more of a need after an acquisition to retain the best practices of the former corporate culture which often is disseminated through their training and development capabilities and staff. Too often, the process employees and managers must go through following a corporate merger or acquisition is not understood by top executive management which leads to unfinished corporate acquisitions.

PLANNING AND RISK ASSESSMENT

There's an old saying about the five P's. Proper Planning Prevents Poor Performance. This should be kept in mind as you consider development of a web based corporate institute for the corporation. As with any new venture there are risks involved. This risk assessment involves considering what you don't know, potential benefits, costs, and how significant the formation of a corporate institute is to the corporation. We have learned to live in a business environment laden with uncertainty. The risks of not venturing forth with new ideas are often greater than seeking out and leading new initiatives.

BENEFITS

A web based corporate institute is an asset to the corporation. Here are some, but not all of the potential benefits in cost savings and cost avoidance if a corporation develops a web based corporate institute:

- Student labor costs can be reduced. Anytime, anywhere learning capability often means when employees have the time and inclination to want to learn. Often that is off company time, during business travel, or on breaks. Time away from the workstation is reduced.

- Facility costs can be reduced. The cost of classroom space, materials storage space, and instructor office space can be reduced by providing an on line corporate institute. In many cases, these costs can be eliminated.

- Capital equipment costs can be reduced. With reduced or eliminated classrooms, there is reduced need for items like overhead projectors, LCD projectors, and other classroom support capital equipment.

- Training staff costs can be reduced. Smaller, but better skilled training staffs are another cost savings to the company. Delivering more instruction with less staff is often the result of optimizing web-based and distance learning technologies.

- Following corporate acquisitions or mergers, a corporate institute can be a catalyst for integrating the best practices of each organization and orienting and training employees to take advantage of the best of both corporate cultures.

It's my belief that customers, business partners, managers, and employees will view the corporate creation of a web based institute as a significant step forward into the future. When I was the Director of The TASC Institute and later the Director of The Veridian Institute I was amazed at the support provided by employees who wanted the institutes developed and implemented so they could continue their self development under the umbrella of a corporate commitment to helping them. The institutes were very successful and as senior management grew accustomed to seeing the institutes managed in a business-like way and as they received positive feedback from their employees and managers on the quality of the education experience, they, too, were inspired to speak favorably about their corporate institutes. As new businesses were acquired by both TASC and Veridian, the corporate institutes helped train thousands of employees and managers as well as provide support for best practices and policies of each corporate culture.

SUMMARY

"Web Based Corporate Institutes; A Solution for Unfinished Defense Corporation Acquisitions" offers the premise that a web based corporate institute can help management to better integrate acquired business units into the corporation and that information can now be made available on the company's intranet and Internet web pages to provide customers, business partners, and employees with access to knowledge, information, and learning opportunities. Since the Internet is a subset of the World Wide Web, I've defined this corporate asset as a web based corporate institute. Most of the labor intensive administrative work now done by training specialists and training administrators can be web based. Employees can be provided with access to learning materials any time, any place. A web based corporate institute can be a valuable asset to the corporation not only by avoiding unnecessary costs and reducing overhead costs for training and development, but also by providing the corporation with an additional profit-making opportunity. Advances in technology require a highly skilled workforce. A corporation's trained employees are the key discriminator in the global marketplace today and any core capabilities of any corporation hinge on the capabilities of the people of that corporation. *In a global marketplace virtually any factor of production, such as capital, raw materials, computers, information technology and so forth can easily be duplicated but what cannot be dupli-*

cated are the knowledge, skills, and attributes of a highly trained and educated workforce.

In the preceding chapters we have explored the challenges corporations face following corporate acquisitions and mergers. Many of the issues are in the area of human resources development and management. We have examined the various missions that can be undertaken by corporate institutes and explored examples of where a corporate institute could report in an organization. We have defined how to develop a web based corporate institute, suggested ways that a corporation could turn an institute into a profit center, and provided examples of corporations that have converted their corporate institutes and universities into profit centers. We've reviewed the lessons learned and difficulties of establishing a corporate institute and the technology trends that support basing the institute on the web. This is a trend that is starting today and will gain momentum in the future.

> *In the future, competitive advantage will accrue to those who invest wisely in developing and retaining human resources, develop web based corporate institutes, and learn how to optimize the new technology available to the corporation.*

▼

GLOSSARY

Bandwidth	The amount of network capacity available to transport and carry files, email, and other data
Distance Learning	A technology enabled method allowing for transmission of instruction to students geographically distant from the source of instruction.
E-Commerce	Conducting business over the Internet and/or World Wide Web.
E-Learning	Learning by taking courses provided over the Internet or World Wide Web.
Firewalls	A technology that secures and protects unlawful entry into a companies intranet.
Internet	A sub set of the World Wide Web that is accessed through Internet Service Providers.
Institute	An organization that focuses on specialized research and training for a specific audience
Intranet	An organization's internal network that can stand alone or or connect to the Internet.
Needs Assessment	A survey of an organizations training requirements
SEI CMM	A five level model developed for software developers to improve the quality of software developed by the Software Engineering Institute and called the Capability Maturity Model.

University   An academic institution made up of several colleges which pro-
                              vides the general public with education

World Wide Web   A global network made up of other networks.

▼

Checklist to Develop a Web Based Corporate Institute

- Be committed to its success. Get "buy-in" at all levels of the organization. Have an executive board of directors, a written policy of governance, and commit the funding and resources needed.

- Understand the training and development requirements of the organization and develop a plan to implement the institute to optimize corporate value.

- Know what is being spent today on training and development, how it's being spent, who's spending the budget allocated, and what the company is gaining from the expenditure. Decide for yourself if a centralized budget will reduce your costs.

- Understand the capabilities and capacity of the company's intranet. Determine what will be Internet based and what will be intranet based.

- Determine the mission, role, reporting relationship, and management of the corporate institute. Explore options and select the model that "fits" your organization and culture.

- Follow a systems engineering approach to developing the institute. Understand the organization's requirements, design a solution, develop the institute, test the model in the organization, and make sure there's a focus on quality and continuous improvement.

- Develop a web-page for the corporate institute and migrate as much information as possible onto the web site. Allow for access across all locations inside the corporation and if the institute has a customer education mission, allow for customers to access the institute from the Internet and the World Wide Web. Develop a communications and marketing strategy for the institute. Make provisions to provide the institute with a logo and brand.

- Constantly benchmark, review, and improve the quality of all of the institutes programs.

About the Author

 **T. H. Henning**, author of *Memoirs of a Defense Contractor*, worked for IBM, Loral, Lockheed Martin, Litton Industries/TASC, and Veridian. As a defense contractor he managed military and government contracts for much of his career and developed web based corporate institutes. He is a veteran of the U.S. Air Force and graduated with a Bachelors and Masters degree from The Pennsylvania State University. He and his wife Cheryl reside in Northern Virginia.

Index

978-0-595-38883-7
0-595-38883-3

www.ingramcontent.com/pod-product-compliance
Lightning Source LLC
Chambersburg PA
CBHW030813180526
45163CB00003B/1264